2nd International Conference on Civil Engineering and Architecture (ICCEA 2019)

IOP Conference Series: Materials Science and Engineering
Volume 690

Seoul, South Korea
21 - 23 September 2019

ISBN: 978-1-7138-0982-1
ISSN: 1757-8981

Printed from e-media with permission by:

Curran Associates, Inc.
57 Morehouse Lane
Red Hook, NY 12571

Some format issues inherent in the e-media version may also appear in this print version.

Institute of Physics
Dirac House, Temple Back
Bristol BS1 6BE UK

Phone: 44 1 17 929 7481
Fax: 44 1 17 920 0979

techtracking@iop.org

Additional copies of this publication are available from:

Curran Associates, Inc.
57 Morehouse Lane
Red Hook, NY 12571 USA
Phone: 845-758-0400
Fax: 845-758-2633
Email: curran@proceedings.com
Web: www.proceedings.com

TABLE OF CONTENTS

URBAN INFRASTRUCTURE CONSTRUCTION AND DEVELOPMENT

2019 2nd International Conference on Civil Engineering and Architecture IOP Publishing
IOP Conf. Series: Materials Science and Engineering **690** (2019) 011001 doi:10.1088/1757-899X/690/1/011001

Preface

Dear Authors and Participants of ICCEA 2019,

On behalf of the conference committees, I am pleased to declare that we had really successful three days of 2019 2nd International Conference on Civil Engineering and Architecture (ICCEA 2019), which was held jointly in Seoul, South Korea during September 21-23, 2019, sponsored by Seoul National University. Again, the conference was of great success, totally beyond my expectation, having a diverse group of expertise and backgrounds. Not to mention that participants are geographically diverse. I was glad to confirm that such diverse expertise and participants were somehow connected, forming a coherent conference environment and creating synergistic learning and sharing.

ICCEA is the premier forum for the presentation of new advances and research results in the fields of theoretical, experimental, and practical civil engineering and architecture. The conference indeed brought together leading researchers, engineers and architects in the domain of interest from around the world.

ICCEA 2019 were composed of 6 oral parallel sessions, 4 keynote speeches delivered respectively by Prof. Ashraf El Damatty, The University of Western Ontario, Canada; Prof. Youngjin Lee, Boston Architectural College & Sasaki Associates, Inc., USA; Prof. T.C. Pong, Hong Kong University of Science and Technology, Hong Kong; and Prof. Patrick Safran, Incheon National University, South Korea; and finally, I would like to thank you to all of our conference committees and participants for always being supportive to the conferences and coming to Seoul during your busy schedule. As expected, the conference was proven to be intellectually stimulating to all of us.

Hope all enjoyed the conference, the food, the hospitality, as well as the beautiful and charming environment of Seoul. I look forward to seeing you again in Beijing in Sept. 2020!

Oct. 1, 2019

Dr. Thomas Kang
Professor, Seoul National University, South Korea
Conference Chair, ICCEA 2019
tkang@snu.ac.kr

Conference Committee

Conference Chair

Thomas Kang, Seoul National University, South Korea

Program Chairs

Ashraf A. El Damatty, University of Western Ontario, Canada
Youngjin Lee, Boston Architectural College & Sasaki Associates, Inc., USA

Technical Committees

Ryan Ahn, Texas A&M University, USA
Yong Han Ahn, Hanyang University, South Korea
Mohammed Alajmi, Kuwait University, Kuwait
Majid Ali, Capital University of Science and Technology Islamabad, Pakistan
Hyeon-Jong Hwang, Hunan University, China
Seung Yup Jang, Korea National University of Transportation, South Korea
Daranee Jareemit, Thammasat University, Thailand
Muneerah Jeludin, Universiti Teknologi Brunei, Brunei
Mohammad Arif Kamal, Aligarh Muslim University, India
Jong R. Kim, Nazarbayev University, Kazakhstan
Sunkook Kim, Kyung Hee University, South Korea
Woosuk Kim, Kumoh National Institute of Technology, South Korea
Yun-Yong Kim, Chungnam National University, South Korea
Sashi Kunnath, UC Davis, USA
Seung-Jun Kwon, Hannam University, South Korea
Hung-Jen Lee, National Yunlin University of Science and Technology, Taiwan
Kang Seok Lee, Hanyang University, South Korea
Leonardo Massone, University of Chile, Chile
Ilwoo Nam, Nanjing Tech University, China
Zhmagul Nuguzhinov, Karaganda State Technical University, Kazakhstan
Yu-Chen Ou, National Taiwan University, Taiwan
Dario Pedrabissi, University of Portsmouth, UK
Alberto G. Perera, Architect & Planner, Spain
Chaiwat Riratanaphong, Thammasat University, Thailand
Samaila Saleh, Hassan Usman Katsina Polytechnic, Nigeria
Junwon Seo, South Dakota State University, USA
Shideh Shadravan, University of Oklahoma, USA
Kyung-Jae Shin, Kyungpook National University, South Korea
Manat Srivanit, Thammasat University, Thailand
Varodom Suksawasdi, Thammasat university, Thailand
Ian F. Taberner, Boston Architectural College, USA
Xiang Guo Wu, Harbin Institute of Technology, China

IOP Conf. Series: Materials Science and Engineering **690** (2019) 011002 doi:10.1088/1757-899X/690/1/011002

Haryati Yaacob, Universiti Teknologi Malaysia, Malaysia
Junyan Yang, Southeast University, China
Yanto Yanto, Jenderal Soedirman University, Indonesia
Wardah Fatimah Mohammad Yusoff, National University of Malaysia, Malaysia

Peer review statement

All papers published in this volume of *IOP Conference Series: Materials Science and Engineering* have been peer reviewed through processes administered by the proceedings Editors. Reviews were conducted by expert referees to the professional and scientific standards expected of a proceedings journal published by IOP Publishing.

Utilization of Recycled Concrete Aggregate for High Performance alkali Activated Concrete: Towards a Sustainable Building Solution

C L Hwang [1], M D Yehualaw[1,2] and D H Vo[1,3]
[1]Department of Civil and Construction Engineering, National Taiwan University of Science and Technology, Taipei, Taiwan
[2] Faculty of Civil and Water Resource Engineering, Bahir Dar Institute of Technology, Bahir Dar University, Bahir Dar, Ethiopia
3 Faculty of Civil Engineering, University of Technology and Education, the University of Danang, Danang, Vietnam

Email: mtkdmt2007@gmail.com

Abstract. The purpose of this study is to prepare environmentally friendly and highly sustainable concrete for an ever-increasing demand of concrete in the construction industry. In the study both natural course and fine aggregates were partially replaced by recycled concrete aggregate from construction and demolition wastes. High performance alkali activated concrete was prepared by utilizing pozzolanic industrial wastes as a binder which fully replaced cement powder. High volume of fly ash with 30%, 40% and 50% replacement of slag was used as a main binder. NaOH and Na2SiO3 with an alkali modulus of 0.6 were used as an alkali activator solution. The findings of this research revealed that alkali activated recycled aggregate concrete have competitive engineering properties with respect to cement concrete with natural aggregate. The UPV and thermal conductivity test results were found to be 4302 m/s and 1.425 W· m-1·K-1 respectively. The results of this study will pave the way towards the better practice of sustainable concrete construction by replacing the cement and natural aggregate in one concrete mix.

1. Introduction

Nowadays, with accelerated industrialization and urbanization increases the rate of industrialization and urbanization has made the use of concrete as the most non-sustainable material. It uses approximately 20 billion tons of raw materials (coarse aggregate) each year. However, it was forecasted that the global consumption of aggregate used in construction may exceed 26 billion tons by 2012. With this increase in rate of consumption, it is expected that the demand of aggregates will be doubled in the next two to three decades [1,2].

Thus, the concrete industry consumes a large amount of natural resources that cause substantial environmental, energy and economic losses as it exploits 50% raw material, 40% of total energy, as well as generates 50% of total waste [2]. So, minimizing the environmental impact, energy consumption and the increase in CO2 intensity for the concrete to be used for construction have become more evident for construction industry which can lead towards sustainable development. On the other hand, Portland cement (PC) manufacturing for concrete work is an energy-intensive process that requires high temperatures (1450-1550 °C), consumes natural resources (mainly limestone and clay) and generates vast greenhouse gas emissions to the atmosphere (approximately 1 ton of CO2 per ton of produced cement. As reported by Torgal et al [3], it is one of the primary causes of global warming, and accounts for 7% of worldwide CO2 emissions.

To alleviate the problem, researchers draw two solutions. The first one is recycling Construction and Demolition Wastes (CDW) as an aggregate for new concrete production. Ajdukiewicz and Kliszczewicz [4] studied the effects of the incorporation of fine and coarse recycled aggregates (RFA and RCA) from demolished concrete with medium/high strength (35-70 MPa) on the performance of concrete. The results showed that using RFA and RCA decreases the compressive strength of concrete from 5-20%. Different researchers improved the performance of Recycle Aggregate Concrete (RAC) by applying chemical admixtures and pozzolanic materials. The second solution is using Alkali activated binder to replace cement. This system pozzolanic materials as a binding material and uses different alkaline solutions such as NaOH and Na2SiO3 as activators. As pointed out by Behera et al [5], the alkali-activation of pozzolanic materials helps to minimize the exploitation of non-renewable raw materials, and to reduce pollution, energy consumption and the areas employed to dispose waste, all of which help mitigate global warming. Alkali activated mortars made by commercial water glass reduce CO2 emissions by 13% compared to PC mortars.

This paper is dedicated to study sustainable and high-performance concrete with utilization of CDW as both RFA and RCA to replace Natural Coarse and Fine Aggregate (NCA, NFA). Cement is fully replaced by slag and fly ash with the application of NaOH and Na2SiO3 as alkaline activators.

2. Material characterization and nature of the concrete mixes

For this study there are two groups of concrete mixes. The first group is the control group, which is made from natural fine and coarse aggregate whereby cement is used as a binder material with w/c ratio of 0.42.

The second group is made from a combination of natural and recycled CDW. The coarse aggregate is composed of 60% NCA and 40% RCA while the fine aggregate is made from 30% RFA and 70% NFA. Recycling CDW replaced 40% and 30% by volume of NCA and NFA respectively. For this group alkali activated fly ash and slag paste is used as a binder with a w/b of 0.38. Three concrete mixes were prepared (F30, F40, & F50) with varying in content of fly ash and slag. The slag was replaced by 30%, 40% and 50% fly ash and the effect on fresh and harden properties of concrete are studied. NaOH and Na2SiO3 are used as alkali activator solutions with M+ and AM value of 4% and 0.6 respectively.

3. Results and discussion

3.1. Workability

The workability of fresh concrete was studied by conducting slump test according to ASTM C 143. The test result is portrayed on Figure 1. It can be understood that the workability of concrete increases with an increase in fly ash and a decrease in slag content. This can be explained that the spherical and glassy nature of fly ash decreases friction [6]. The control mix has higher workability than F30 and lower than F40 & F50. To illustrate figuratively, F30 and F50 concrete mixes showed the least and the highest slump value of 220 and 195mm respectively while the control mix and F40 showed similar slump value of 210mm. All concrete mixes have very good workability with very close figure with a range of 25mm.

IOP Conf. Series: Materials Science and Engineering **690** (2019) 012001 doi:10.1088/1757-899X/690/1/012001

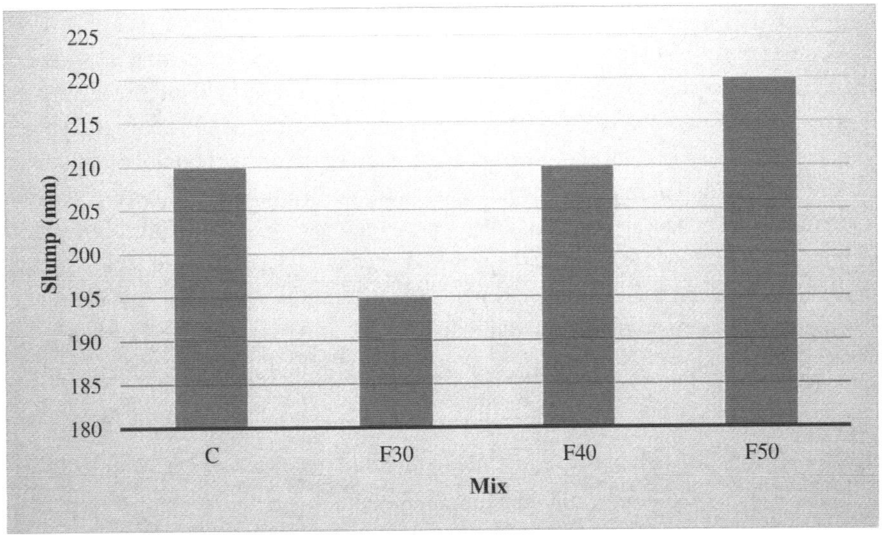

Figure 1. Workability of concrete mixes

3.2. Compressive strength

The compressive strength results of concrete mixes were tested on 3, 7- and 28-days age according to ASTM C 109 and the results are summarized on Figure 2. It can easily understand that the compressive strength of concrete mixes with recycled coarse and fine aggregate is lower than the concrete mix with natural aggregate. This is due to the adhered mortar on the surface of recycled course and fine aggregate hinder to make strong bond between aggregate and alkali activated paste. This explanation agreed with previous research works [7]. The compressive strength of recycled aggregate concrete mixes was improved with an increase in content of slag. This is actually due to high reactivity nature of slag as can be seen on Figure 1. The control concrete mix showed the highest strength at 28 day with 60MPa while F30, F40 & F50 showed 51,46 & 40MPa respectively.

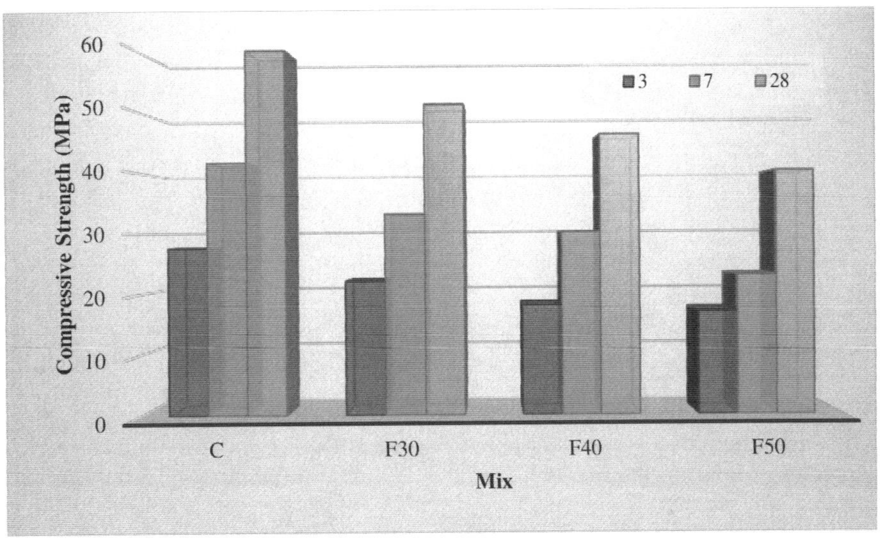

Figure 2. Compressive strength of concrete mixes

IOP Conf. Series: Materials Science and Engineering **690** (2019) 012001 doi:10.1088/1757-899X/690/1/012001

3.3. Ultrasonic Pulse Velocity (UPV)

Different researchers have demonstrated UPV test to measure the durability of concrete such as porosity and permeability value to assess the compressive strength and durability of properties of concrete [8]. The UPV test results of control and alkali activated concrete mixes is displayed on table 1. It can be understood that UPV results drops with in increase in fly ash content. The UPV results of control and alkali activated concrete mixes corresponds to their respective compressive strength results. This is also supported by research result of different researchers [9]. The control mix revealed UPV value of 4525m/s at 28 days age while F50 displayed 4265m/s. All alkali activated recycled aggregate concrete group mixes showed lower results than the control mixes due to the low density of RFA and RCA. According to the suggested UPV ratings of Malhotra et al. [10], all concrete samples are considered to be good quality concretes, because all UPV results are in the range of 3660–4575 m/s

Table 1. UPV and thermal conductivity test results of concrete on 28th date age

Mix	UPV (m/s)	Thermal Conductivity (W/m*k)
C	4525	1.455
F3	4302	1.425
F4	4294	1.335
F5	4265	1.29

3.4. Thermal conductivity

The ability of the concrete to absorb heat is measured by applying thermal conductivity test. The result of thermal conductivity test on 28th date is summarized on Table 1. It was found that the thermal conductivity of concrete made from alkali activated recycled aggregate is lower than the control concrete with natural aggregate. This is basically due to the high porosity of RFA and RCA. Since the conductivity of air is very low therefore, conductivity of Recycled Aggregate Concrete (RAC) varies with the variation in aggregate density or porosity [11]. Moreover, since cement has higher thermal conductivity than fly ash and slag thermal conductivity of control concrete mix is higher than Alkali activated RAC. However, the thermal conductivity of alkali activated RAC improved with an increase in slag content due to its very amorphous nature it produces more alkali activated paste and makes the concrete micro structure of concrete to be denser [12,13].

4. Conclusion

From the results of the study the following points can be concluded:

- Recycled CDW can be simultaneously used as coarse and fine aggregate for the development of RAC for a high-performance concrete
- The compressive strength of alkali activated RAC is lower than the control concrete. F50% has 33% lower strength than the control mix.
- UPV and thermal conductivity show similar trends with the compressive strength of concrete mixes. They improve with an increase in compressive strength of concrete.
- Alkali activation system is the best alternative to make sustainable concrete by replacing the environmentally unfriendly, costly and energy intensive ordinary Portland cement.

5. References

[1] Sonawane T, Pimplikar S., Use of recycled aggregate in concrete. Int J Eng Res Technol ;2(1):1–9. 2013.

[2] Oikonomou N , Recycled concrete aggregates. Cem Concr Compos; 27:315–8 2005.

[3] Torgal P., Labrincha J, Leonelli C., Palomo A., Chindaprasit P., Handbook of Alkali-Activated Cements, Mortars and Concretes1st edition, Woodhead Publishing, United Kingdom, 2014.

[4] Ajdukiewicz A., Kliszczewicz A., Influence of recycled aggregates on mechanical properties of HS/HPC. Cem. Concr. Compos 24 (2), 269-279, 2002.

[5] Behera M., Bhattacharyya S, Minocha A, Deoliya R, Maiti S, Recycled aggregate from C & D waste & its use in concrete - A breakthrough towards sustainability in construction sector: a review, Constr. Build. Mater. 68 501–516 2014.

[6] Hwang C, Yehualaw M, Vo H, Huynh P, development of high strength alkali activated paste containing high volume of waste brick and ceramic powders, construction and building materials, 218 (2019) 519-529.

[7] Pedro D, Brito D, Evangelista L, Structural concrete with simultaneous incorporation of fine and coarse recycled concrete aggregates: Mechanical, durability and long-term properties, Construction and Building Materials, 154 294-309 2017.

[8] Lafhaj Z, Goueygou M, Djerbi A, Kaczmarek M, Correlation between porosity, permeability and ultrasonic parameters of mortar with variable water/cement ratio and water content. Cem. Concr. Res., 36(4), 625–633 2006.

[9] Hwang C, Tran V, Engineering and Durability Properties of Self-Consolidating Concrete Incorporating Foamed Lightweight Aggregate, J. Mater. Civ. Eng., 28(9) 2016.

[10] Malhotra V, Testing Hardened Concrete: Nondestructive Methods American Concrete Institute Monograph No. 9, Detroit, Michigan, 1976.

[11] Chung S, Han T, Kim S, Kim J, Youm K, Lim J, Evaluation of effect of glass beads on thermal conductivity of insulating concrete using micro CT images and probability functions, Cem. Concr. Compos. 65 150–162 2016.

[12] Demirboğa R., Influence of mineral admixtures on thermal conductivity and compressive strength of mortar, Energy Build. 35 (2) 189–192 2003.

[13] Bentz D, Peltz, Duran M ,Valdez A., Juárez P, Thermal properties of high-volume fly ash mortars and concretes, J. Build. Phys. 34 (3) 263–275 2011.

IOP Conf. Series: Materials Science and Engineering **690** (2019) 012002 doi:10.1088/1757-899X/690/1/012002

Assessment of Bending Reinforced Concrete Beams Crack Resistance

Zh S Nuguzhinov[1], Zh B Bakirov[1], A Yu Kurokhtin[1], O Khabidolda[2] and A Nuguzhinova[1]

[1]Kazakhstan Multidisciplinary Institute of Reconstruction and Development at KSTU, Karaganda, Kazakhstan
[2]Mechanics Department, Mechanics and Mathematics Faculty, Al-Farabi Kazakh National University, Almaty, Kazakhstan.

Email: kazmirr@mail.ru

Abstract. The problem of determining the stress state in a section of bending reinforced concrete beams with an initial and operational cracks has been solved analytically. To this end, a system of two nonlinear algebraic equations has been obtained from the equilibrium conditions of the part of the beam cut along the crack line. From this system, for a beam with an initial crack, the height of the compression zone and the nominal stress at the crack tip are determined, and for an operational crack, the height of the compression zone and the length of the crack are determined. The remaining parameters of the stress state are expressed in terms of these values. There is also determined the value of the external moment above which an increase in the initial length of the crack occurs.

Determining the stress intensity factor (SIF) is based on the assumptions that the longitudinal forces are equal at the crack tip, with and without stress concentration. The size of the stress concentration zone is determined from the condition that the local stress is equal to the nominal stress. On this basis, the formula for determining the SIF is obtained. The paper analyzes the SIF dependence on the crack length and the bending moment.

The method of calculation is valid for beams of arbitrary cross-section, but explicit dependences are given for beams of rectangular cross-section most frequently encountered in production.

The results obtained allow estimating the bearing capacity of beams with a crack, as well as their crack resistance according to the force criterion of fracture mechanics.

1. Introduction

Cracks are a very common type of defect in reinforced concrete structures. They occur at the manufacturing stage and at the operation stage. The appearance of cracks in bending elements does not mean exhaustion of its bearing capacity. It leads to increasing the internal efforts in the sections with a crack, which reduces the element strength. Due to the opening of the crack width, corrosion of reinforcement increases, which reduces the durability of structures. A lot of books [1, 2] and articles [3-7] are dealing with calculating the stress state of reinforced concrete structures with cracks. In these works, the length of the crack is considered known. However, during bending the length of the crack depends on the bending moment, and the length of the existing crack may increase. The crack opening width is one of the criteria for the ultimate state of reinforced concrete structures with cracks. Works

[8–10] are dealing with its definition. The issues of crack formation are also considered in works [11, 12].

Under certain conditions, unstable crack development is possible, which is estimated in terms of the parameters of fracture mechanics [13, 14]. The issues of applying fracture mechanics criteria are still under discussion [15, 16]. The most frequently used stress intensity factor (SIF) is the main parameter of linear fracture mechanics. In the general case, the SIF is determined by the finite element method in the linear formulation. However, the relationship between stress and strain in concrete is not linear. This introduces inaccuracies in the SIF definition in a reinforced concrete element. During bending, the development of a crack is limited to reinforcement in the tension zone and is blocked by the presence of a compression zone, which introduces its own characteristics in the SIF definition.

In this paper, the authors propose an approximate analytical method for determining the SIF in reinforced concrete beams. For this, the stress state in a beam with a crack is first determined. The authors assume that during bending the crack starts from the contour point of the tension zone that is the farthest from the neutral axis and spreads across the cross section.

2. Determining stress state

Let's consider the case of bending an I-beam with a pre-stressed lower reinforcement and a crack (see Figure 1,a). When deriving the Formulas, we assume that the law of flat sections is satisfied. Two cases must be distinguished here: a) the nominal stress at the crack tip is lower than or equal to the ultimate tensile strength of concrete b) exceeds which leads to increasing the initial crack length l. When calculating the stress state, we consider the law of flat sections to be fair. Here it is necessary to distinguish two cases: a) the nominal stress at the crack tip σ_m is lower than or equal to the tensile strength of concrete R_{bt}; b) this stress exceeds R_{bt} which leads to increasing the initial crack length l.

Let's consider an I-section with the vertical axis of symmetry (Fig. 1, a). Let's introduce designations:

A_{ct}, A_c is the area of the shelves overhang in the tension and compression zones;

h_t, h_c is the thickness of the shelves in the tension and compression zones;

a, a' is the thickness of the protective layer of concrete in the tension and compression zones;

A_s, A_s' is the area of reinforcement in the tension and compression zones;

N_a, N_a' are internal forces in the reinforcement in the tension and compression zones.

Let's cut the beam by the section passing through the crack and show the curve of the stress distribution in it.

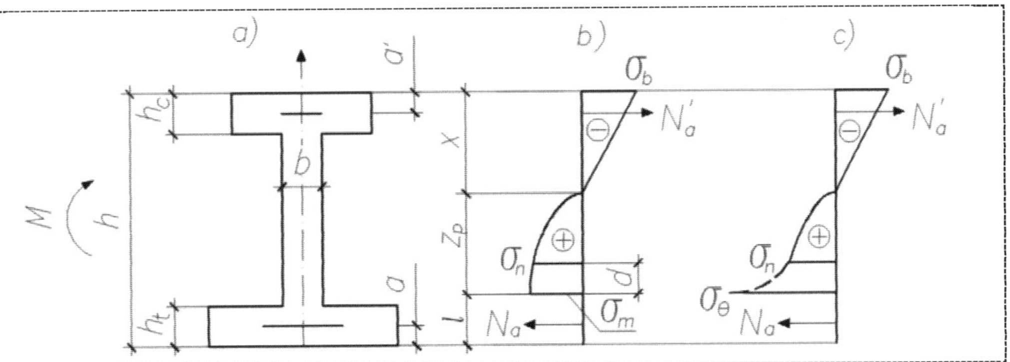

Figure 1. Diagram of stress distribution in the cross section of the bending pre-stressed bean with a crack: (a) – cross section, b) – without considering the stress concentration at the crack tip, c) – taking into account the stress concentration at the crack tip

For the tension zone the relationship between stress and strain has the form of the exponential law [17].

$$\sigma = 1.1R_{bt}\left[1-\exp(-0.9\varepsilon E_b / R_{bt}\right], \quad (1)$$

IOP Conf. Series: Materials Science and Engineering **690** (2019) 012002 doi:10.1088/1757-899X/690/1/012002

where E_b and R_{bt} are the elasticity modulus and the concrete ultimate strength.

Due to the low stress level ($\sigma_m<R_{bt}$), the distribution of compressive stresses is assumed to be linear. Then the stress profile in the cross section will take the form shown in Figure 1, b.

Tensile stress at the tip of the crack σ_m and the height of the compression zone x. are taken as the basic unknown values of the problem. We introduce designations:

$$c=\sigma_m/1.1R_{bt}, \qquad y=-\ln(1-c) \tag{2}$$

Then the strain at the crack tip is determined by the expression:

$$\varepsilon_m=1.1yR_{bt}/E_b.$$

The strain and stress at the edge of the compression zone taking into account the hypothesis of flat sections will be:

$$\varepsilon_b=\varepsilon_m x/z_p, \quad \sigma_b=E_b\varepsilon_b=1.1yR_{bt}x/z_p, \tag{3}$$

where z_p is the height of the tension zone.

In the zone of the crack, the adhesion of the reinforcement to the concrete is broken, the tensile forces are mainly perceived by the reinforcement. In the sections between cracks the tensile forces are perceived by reinforcement and concrete. The strains and stresses in the reinforcement and concrete, as well as the height of the compression zone between the cracks vary. This non-uniformity in the calculations is taken into account by introducing special coefficients that are equal to the ratios of the average values in the section between the cracks and the values in the section with a crack [18]

$$\psi_s=\varepsilon_{sm}/\varepsilon_s, \quad \psi_b=\varepsilon_{bm}/\varepsilon_b.$$

Taking into account the strain and stresses in the reinforcement will be:

$$\begin{aligned}
&\varepsilon_s=\varepsilon_m\psi_{bs}(h-a-x)/z_p,\\
&\sigma_s=E_s\varepsilon_s=1.1yR_{bt}\alpha\psi_{bs}(h-a-x)/z_p,\\
&\sigma'_s=\sigma'_{sp}-E_s\varepsilon'_s=\sigma'_{sp}-1.1\alpha yR_{bt}(x-a')/z_p,
\end{aligned} \tag{4}$$

where α is the ratio of the reinforcement and concrete elasticity modulus; $\psi_{bs}=\psi_b/\psi_s$; σ'_{sp} is pre-stress in the compression zone reinforcement.

The tensile stress at the distance z from the zero line, taking into account (1) and the flat section hypothesis, has the form:

$$\sigma=1.1R_{bt}[1-\exp(-yz/z_p)]. \tag{5}$$

The resultant of the internal forces in the tension zone and their moment relative to the zero line are determined by integrating over the height of the tension zone

$$N_p=nb\sigma_m z_p, \quad M_{po}=m\sigma_m bz_p^2, \quad M_{pc}=M_{po}-N_p(h/2-x),$$

where $n=1/c-1/y, \quad m=[0.5+(1-c)(1+y)/y^2-1/y^2]/c;$

M_{pc} is the moment relative to the central axis.

The condition of the beam cut-off zone equilibrium as the sum of the projection of forces on the beam axis is:

$$N_p+\sigma_s A_s-\sigma_b bx/2+\sigma'_s A'_s-\sigma_b A_c(x-h_c/2)/x=0.$$

The second equilibrium equation as a sum of moments of all the forces relative to the concrete central axis has the form:

$$M_{pc}+\sigma_s A_s(h/2-a)+0.5\sigma_b bx(h/2-x/3)+\sigma_b A_c(x-h_c/2)(h-h_c)/2x-\sigma'_s A'_s(h/2-a')+1.1y\alpha R_{bt}A_s'(x-a')(h/2-a')=M_{bn},$$

where the external moment M_{bn} is equal to the bending moment M in the section with a crack for the reinforcement without pre-stress. For the pre-stressed reinforcement of the tension zone, stress σ_s occurs after the moment of external forces M exceeds the moment of the pretension force. Then in this equation the total external moment will be equal to

$$M_{bn}=M-\sigma_{sp}A_s(h/2-a).$$

The equilibrium equations are common for all reinforced concrete elements with and without pre-stressing, with various cross-sectional shapes: I-beams, T-shaped, rectangular. For a T-section A_c or A_{ct} is equal to zero. For a rectangular section the areas of both overhangs are zero. In order to obtain explicit calculated ratios, we will carry out a further solution for the most common case of the rectangular section with reinforcement in the tension zone.

Let's introduce the following dimensionless parameters:

$$\xi = x/h,\quad z = l/h,\quad \lambda = z_p/h = 1 - z - \xi,\quad \bar{h} = (h-a)/h,\quad \mu = A_s/bh.$$

Then, taking into account the expressions for σ_b and σ_s the equilibrium equations can be written in the dimensionless form:

$$nc\lambda^2 - 0.5y\xi^2 + y\alpha\mu\psi_{bs}(\bar{h}-\xi) = 0. \tag{6}$$

$$L = M_{bn}/bh^2R_{bt} = c\lambda\left[m\lambda - n(0.5-\xi)\right] + y\alpha\mu\psi_{bs}\left(\bar{h}-\xi\right)(0.5-a_0)/\lambda + 0.5y(0.5-\xi/3)\xi^2/\lambda.$$

The solution of this system of nonlinear equations makes it possible to determine the stress at the crack tip and the height of the compression zone depending on the external moment.

This solution is valid as long as the nominal stress is lower than or equal to the tensile strength of the concrete R_{bt}. Otherwise, the initial crack length will increase. The value of the moment M_m at which this happens is determined from system (6), if we take $\sigma_m = R_{bt}$. At this $c = 0.909,\ y = 2.4,\ n = 0.683,\ m = 0.418$.

The new crack length corresponding to the effective moment $M > M_m$, is also determined from system (6) at $\sigma_m = R_{bt}$. To determine stresses by formulas (3) and (4) and to calculate the SIF, it is needed to determine ξ and λ from this system.

3. Determining stress intensity coefficient

Let's determine the SIF at the tip of the crack K_I. The tip of the crack is a strong stress concentrator. According to the linear theory of elasticity, stress there is determined by the

$$\sigma_\theta = K_I/\sqrt{2\pi r}, \tag{7}$$

where r is the distance from the crack tip to the point under consideration.

The SIF determining is based on the equality of the longitudinal forces at the crack tip with and without stress concentration. Local stress is determined by formula (7), and the nominal stress by formula (5).

Let's determine the length of the stress concentration zone "*d*" from the condition of equality of the n rated and local stress at the end of this zone (Figure 1,c). Rated stresses at the distance "*d*" from the crack tip

$$\sigma_n = 1.1R_{bt}\left[1-(1-c)^{1-t}\right],\quad t = a/z_p.$$

The longitudinal forces per unit width of the beam in the zone of concentration are

$$I_1 = \int_0^a \sigma_\theta dr = K_I\sqrt{2a/\pi} = K_I^2/\pi\sigma_n.$$

The longitudinal forces in the same zone without stress concentration are equal

$$I_2 = 1.1R_{bt}z_p\left\{t + \left[1-c-(1-c)^{1-t}\right]/y\right\}.$$

By equating these forces, we obtain

$$f(t) = t - (1-c)^{1-t}(2t - 1/y) - (1-c)/y = 0.$$

Solving this equation with the known $c(y)$, we find the parameter *t*.

Then from (7) we determine the SIF:

$$K_I = \sqrt{2\pi a}\cdot\sigma_n = \sqrt{2\pi}\cdot 1.1R_{bt}\left[1-(1-c)^{1-t}\right]\sqrt{t\lambda h}. \tag{8}$$

If the external moment is larger than M_m, then $\sigma_m = R_{bt}$, $c = 0.909$, $t = 0.808$ and

$$K_I = 0.914 R_{bt} \sqrt{z_p}.$$

The calculations show that with increasing the crack length, the SIF decreases to zero. This is explained by the fact that during bending there is necessarily a minimum zone of compressive stresses. With increasing the crack length, this reduces the length of the zone of tensile stresses to zero. At $z_p = 0$, tensile forces are perceived only by reinforcement, the crack does not develop. In this case, the SIF becomes zero and the beam is destroyed due to the achievement of the yield strength in the reinforcement or crushing of the concrete of the compression zone.

The paper analyzes changing the SIF with changing the external moment. The calculations have been carried out for a rectangular cross-section beam of size (15x30) *cm* made of grade B25 concrete (R_{bt}=*1.6 MПa,* R_b=*14.5 MPa*) with reinforcement made of AIII steel (R_s=*370 MПa,* $\alpha\mu\psi_{bs}$=*0.15*). The beam has an initial crack l0=6 *cm*. For this case, the crack growth moment is determined from system (6) and is equal to M_m = *9.96 kNm*. The results of the calculations are shown in the Table 1 below.

Table 1. SIF changing depending on the external moment.

M, kNm	2	6	10	16	26	34	43
K_I, $MN/m^{3/2}$	0.128	0.316	0.467	0.386	0.31	0.25	0.14

It can be seen in the Table that for $M<M_m$, the SIF increases with increase the moment. When $M>M_m$, the crack grows and the SIF decreases with increasing the moment.

If, with a known crack length and a given moment, the SIF is lower than the limit SIF value for a given material, then the crack will develop steadily. If $K_I \geq K_{IC}$, the crack is non-stable and rapid crack growth is possible with a slight increase in the moment. The increase in the length of the cracks continues until the SIF drops to the K_{IC} value. This value of the SIF corresponds to a certain value of the crack length, which is determined from system (6) and expression (8).

Evaluation of the beam bearing capacity is performed by the value of the stress in the reinforcement (4) and the maximum compressive stress in the concrete (3). The parameters σ_m, x, z_p, found through the initial ($M<M_m$) or operational ($M>M_m$) crack length are substituted into these formulas.

4. Conclusions

Based on the results of the work, the following conclusions can be drawn.

1. The parameters of the stress state of bending reinforced concrete beams with an initial and operational crack are determined.
2. The value of the external moment is determined, above which the crack length increases.
3. By analyzing the stress state in the crack zone, the stress intensity factor at the crack tip has been determined.
4. The effect of the crack length and the external moment on the magnitude of the ORF is analyzed. It has been established that in a reinforced concrete beam with increasing crack length, CIN decreases.

The results obtained allow estimating the bearing capacity of reinforced concrete beams with a crack, as well as estimating crack resistance of the beam according to the force criterion of fracture mechanics.

The method of calculation is valid for beams of an arbitrary cross-section, but explicit dependences are given for beams of the ectangular cross-section most frequently encountered in production.

5. References

[1] Nuguzhinov Zh S 2012 *Theory of calculating reinforced concrete structures with damage* (Karaganda: KSTU publishing house) p 203

[2] Kodysh E N, Nikitin I K and Trekin N N 2010 *Calculation of reinforced concrete structures made of heavy concrete for strength, crack resistance and deformations.* (Moscow: ABC Publishing house) p 352
[3] Kolchunov V I and Demyanov A I 2018 *Engineering and Construction Journal* **No** 5(81) pp 160–173
[4] Chernin L and Val D V 2014 *Construction and Building Materials* **No** 4 **V** 25 pp 1854-1869
[5] Tamrazyan A A, Orlova M M 2016 *Scientific Review* **No** 6 pp 8-11
[6] Tamrazyan A A, Orlova M M 2015 *Vestnik TGSAU* **No** 6(53) pp 98-105
[7] Ma F J, Kwan A K H 2018 *Advances in Structural Engineering* **No** 10 **V** 21 pp 1454-1468
[8] Kwan A K H, Ma F J 2016 *Engineering Structures* **V** 126 pp 618-627
[9] Oliver-Leblond C, Delaplace A, Ragueneau F 2015 *Engineering Structures* **V** 83 pp. 176-186
[10] Amin A, Gilbert R I 2018 *ACI Structural Journal* **V** 115 **Issue** 2 pp 535-543
[11] Hua Zhu 2016 *FracturaedIntegrit à Structurale* **No** 36 pp 191-200
[12] Sheng J, Yin S P, Xu S L, Jin Z 2018 *Advances in Structural Engineering* **V** 21 pp 1696-1707
[13] Zaitsev Yu V 1991 *Fracture mechanics for builders* (Moscow: Higher School) p 288
[14] Cherepanov G I 2012 *Fracture mechanics* (Izhevsk: IKI) p 872
[15] Garcia-Alvarez Victor O, Gettu Ravindra Carol Ignacio 2012 *Sadhana-academy proceedings in engineering sciences* **V** 37 pp 187-205
[16] Klyueva N V, Kolchunov V I, Yakovenko N A 2014 *News of Kazan State Architectural and Construction University* **No** 3 pp 41-45
[17] Almazov V O 2011 *Designing reinforced concrete structures according to Eurocodes* (Moscow: DIA Publishing house) p 216
[18] Karpenko N I 1996 *General models of reinforced concrete mechanics* (M.: Stroyizdat) p 416

IOP Conf. Series: Materials Science and Engineering **690** (2019) 012003 doi:10.1088/1757-899X/690/1/012003

A Review on the Seismic Performance of Assembled Steel Frame-precast Concrete Facade Panels

Weiyu Wang[1], Zhenbao Li[1,*], Wenjing Wang[2], Zhiwu Li[3,4] and Hui Wang[1]

[1] The Laboratory of Urban Security and Disaster Engineering of Ministry of Education Beijing University of Technology, Beijing, China
[2] China State Construction Engineering Corporation (CSCEC), Beijing, China
[3] China State Construction Development Co., Ltd., Beijing, China
[4] China Construction Science & Technology Co., Ltd., Beijing, China

Email: wwylearning@163.com

Abstract. Assembled steel structure housing has the advantages of low cost, light weight and flexible space arrangement, which has been largely used in government-subsidized housing in China. Meanwhile with the rising of assembled steel structure housing, precast concrete facade panels have been widely used. In this paper, the research on the seismic performance of assembled steel frame with precast concrete facade panels in China is summarized and discussed, including the influence of precast concrete facade panels, flexible connections and the seismic performance of steel frame influenced by precast concrete facade panels. According to the research status, the mechanical characteristic of the precast concrete facade panels following the deformation of steel frames; the seismic performance influence of the main structure in rare earthquakes because of the flexible connections are analysed. This paper provides a reference for seismic theory establishment of assembled steel structure housing envelope system.

1. Introduction

China is an earthquake-prone country, between the circum-Pacific seismic belt and the Eurasia seismic belt. The market share of steel structure buildings in the civil construction market is far lower than that of industrial buildings and long-span spatial structures market [1]. There are three design principles of assembled buildings: “Modularized” basing on basic module design; “Standardized” basing on less variety, high production efficiency, and more combinations design; “Integrated” basing on whole process and whole system details pre-design [2]. It is imperative to put assembled steel structure housing system into industrialization, because of which can perfectly make building construction more collaborative and integrated. With the development of assembled steel structure housing system, the seismic performance of non-structural components has been attracted more attention.

Assembled steel structure housing system includes structure system, building envelop system and interior decoration system. The building envelop system increases the lateral stiffness of the steel frame structure and participates in the horizontal force. The building envelop system includes interior wall panels, external wall panels, doors, windows and roof panels. The research on the interior wall panels was carried out earlier, mainly discussed about infilled walls [3-7]. The theoretical research on the external wall panels [8] is also in the process of digging.

The installation forms of external panels include embedded building and external hanging. The research on the embedded panels were carried out earlier by foreign researchers, the design of many non-structural components is still based on these earlier foreign research results. However, with the improvement of China's science and technology, the research on the bearing capacity of precast concrete facade panels in recent years is being studied deeply by domestic researchers [9-14]. The calculation method of precast concrete facade panels bearing capacity mainly refers to the current Chinese standard *Code for Design of Concrete Structures*GB50010 [15], *Standard for Design of Steel Structures* GB50017 [16], *Code for Seismic Design of Buildings* GB50011 [17], *Technical Standard for Assembled Buildings with Concrete Structure* GB/T51231 [18] and *Technical Specification for Precast Concrete Structures* JGJ1 [19]. *Technical Standard for Application of Precast Concrete Facade Panels* JGJ/T485 [20], as an industry standard, was promulgated on December 27, 2018.

If precast concrete facade panel designed unproperly, it will easily lead to a secondary disaster, the panel will possibly fall off or even collapse. Hetao H [21-22] and Youdong C [23] pointed out that future research directions of precast concrete facade panels should focus on the following aspects: the improvement of wall panels design theory, the improvement of joint connection forms and the manufacture of new materials. In this paper, aiming to point out the further research direction and approach of seismic performance of precast concrete facade panels system, the research results of seismic performance of assembled steel frame-precast concrete facade panels system are classified and summarized by Chinese experts and scholars in recent years.

2. Precast concrete façade panels

2.1. Design formulas

The calculation of bearing capacity of precast concrete facade panels includes load combination calculation and ultimate bearing capacity checking calculation. Reference [20] clearly stipulates that the load combination should consider the influences of the precast concrete facade panels self-weight, wind load, earthquake action, thermal action and the deformation of the main structure. According to reference [19], the horizontal seismic action calculation of precast concrete facade panels follows the regulation of seismic analysis method for equivalent lateral force, which means that the critical direction of horizontal seismic action is related to influence coefficient of frequent earthquakes, gravity and dynamic magnification coefficient, as shown in formula (1). Guowei Z [12], Jiasen L [13], and Jinhu W [24] calculated wind load and seismic action when carrying out load combination, and seismic action was calculated according to the above method.

$$F_{Ehk} = \beta_E \alpha_{\max} G_k \tag{1}$$

where, F_{Ehk} is the standard value of horizontal seismic action applied to the gravity center of the precast concrete facade panels; β_E is the dynamic amplification factor, value 5.0; $\alpha_{\max}$ is the maximum influence coefficient of horizontal earthquakes. G_k is the standard value of gravity load (kN) of precast concrete facade panels.

According to reference [17], the value of amplification coefficient is related to the function, category, state and floor position of non-structural components, and the horizontal seismic action is shown in formula (2).

$$F = \gamma \eta \xi_1 \xi_2 \alpha_{\max} G \tag{2}$$

where, γ is the function coefficient for non-structural components, for precast concrete facade panels should be bear 1.4 times the design deflection of the main structure components; η is the classification coefficient of non-structural components, for retaining wall η is 0.9; ξ_1 is 2.0 of a prefabricated component as a state factor; ξ_2 is position factor, taking with an interval of 1.0 to 2.0.

2.2. Test and finite element analysis

Hongchao G [25], in order to study the force transfer path of precast recycled concrete facade panels, conducted a pseudo-static test with axial compression on three trusses of one-story one-span full-scale flat model. The steel column height was 1275mm. The seismic performance of the steel frame was evaluated according to the horizontal bearing capacity, lateral stiffness, displacement ductility and energy dissipation capacity. Research showed that the force transferring path of precast recycled concrete facade panels started from the connection and passed along the bottom edge of the diagonal.

Besides the traditional concrete materials, the precast concrete facade panels also derive various materials, such as light weight panels and composite panels. 2 spatial full-scale models of three stories, three span full-scale steel frame-precast concrete facade composite panels was carried on the shaking table test by Wei W [26]. The deformation characteristics of the precast concrete facade composite panel were studied. The precast concrete facade composite panel was composed of ceramist concrete sandwich panel and EPS panels. The working conditions of the model included frequent earthquakes, seismic fortification earthquakes, rare earthquakes and aftershock. Models can be easily repaired and finally accepted by the super-large seismic waves. Research suggested that the main structure of the residual displacement was less than 0.5% when precast concrete facade composite panels could be better to follow the main structure deformation. The precast concrete facade panels had a good ability to adapt to the deformation of the main structure during the test vibration process. On the basis of the experiment in reference [27], Mingji F [28] drew a comparison diagram of the acceleration response of each floor between the floor and the precast ALC facade panels at peak time to prefabricate the acceleration response of the precast ALC facade panels. ALC means Autoclaved Lightweight Concrete. Studies showed that precast ALC facade panels would produce independent acceleration to steel frame. Independent acceleration related to the independent deformation and energy dissipation of flexible connection.

Jianyao G [29] used SAP2000 to analyse the time history response of middle, low and high-rise steel frame structures under different earthquake working conditions respectively. The results had shown that in the rare earthquake, the dynamic response of the precast concrete facade panels to meet twice than the frequent earthquake. Basing on the reference [25], Lijian S [30-31] compared strain curve of the precast recycled concrete facade panels and joint connections. They also found that the deformation coordination performance between steel frame and the precast recycled concrete facade panels was good. It was suggested that the upside connections of precast recycled concrete facade panels should get involved with deformation and energy dissipation, while the downside connections should bear stress.

In the newly issued reference [19], the calculation of horizontal seismic effects of the panels was more clearly specified: the amplification coefficient of the precast concrete facade panels in case of frequent earthquakes was set at 5.0. However, in the actual study for precast concrete facade panels of the seismic response did not reach 5.0 times yet [31]. According to different forms of precast concrete facade panels taking with proper amplification coefficient is still a problem need to be studied further.

3. Prefabricated steel frame-precast concrete facade panels joint connections

3.1. Design formulas

Precast concrete facade panels can be applied to different forms of main structures [32-33]. Flexible connection can ensure that the precast concrete facade panels can dissipate energy through the rotation or displacement by using the joint connections when the main structure deforming. This chapter mainly introduced the design criteria of joint connections and Chinese research progress when the assembled steel frame as the main structure flexibly connected with the precast concrete facade panels.

The connections between precast concrete facade panel and steel frame are divided into flexible connection and rigid connection. Figure 1 and figure 2 in reference [20] is a schematic diagram of the connection form of the precast concrete facade panel. Both of them have different advantages. The flexible connection is a typical point support form, which has the ability to adapt to deformation. The rigid connection is a typical linear support form, which can increase the structural stiffness and participate in bearing force. According to the different roles of joint connections, the point-supported precast concrete facade panel can be divided into displacement type, rotation type and fixation type.

IOP Conf. Series: Materials Science and Engineering **690** (2019) 012003 doi:10.1088/1757-899X/690/1/012003

Regularly, the connections of a precast concrete facade panel should be set on the upside and downside, but the bearing joint and non-bearing joint should not be set on the same side.

When bearing capacity is calculated, it is necessary to consider the stress condition of joint connections under different loads, and the stress form is shown in figure 3 [24]. Self-weight of precast concrete facade panel should be designed at the position of the load-bearing joint; Wind load was equally divided by load-bearing joint and limitation joint [29]. The horizontal seismic action outside the panel made the joint connections under pressure (or tension), and its size is divided equally by load-bearing joint and limitation joint. The horizontal seismic action inside of the precast concrete facade panel made the joint connection bearing shear force, and each load-bearing joint bears the 0.5 times of horizontal seismic action. Each limitation joint bear 0.25 times of the horizontal seismic action. The calculation method of earthquake action of precast concrete facade panel refers to formula (1). According to reference [20], when the dynamic amplification coefficient of the joint connections of precast concrete facade panel were designed depending on the seismic fortification intensity of the region or the rare earthquake, it could be set as 4.0 or 5.6 according to the building type.

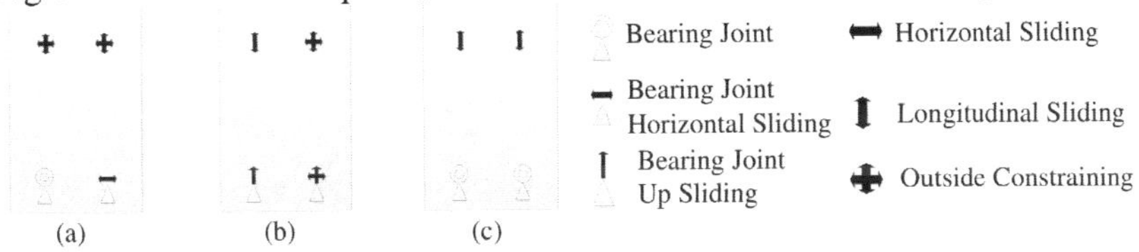

Figure 1. Flexible Connections, displacement type of (a), rotation type of (b) and fixation type of (c).

Figure 2. Rigid Connection.

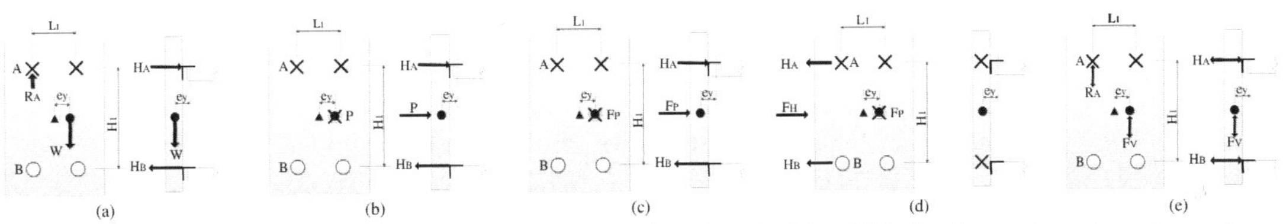

Figure 3. Stress diagram of joint connections: gravity(a), wind load(b), horizontal earthquake action—outside(c), horizontal earthquake action—inside(d), horizontal earthquake action—outside(e).

3.2. Test and finite element analysis

Guoqiang L [28] [34-35] deduced the maximum shear stress of joint connections based on the inertia force of precast concrete facade panel from the shaking table test data. It was found that the moment and bending stress will be regenerated by the shear force when the joint connection was under stressed. It was suggested that the non-uniform coefficient of stress on precast concrete composite external wall panel should be considered in the joint design. After the 9-degree earthquake, the joint connection between the precast ALC facade panel and the steel frame was subjected to the inertia force, which was determined by the mass and acceleration of the ALC facade panel. The full-scale shaking table test carried out by Wei W [26]. The test results showed that under the action of super-large earthquake, the damage of the ALC facade panel was the maximum damage at the joint connection and embedded parts.

Yong J [36-37] summarized and tested the bearing capacity of four types of flexible joint connections. The upside load-bearing type joint meant the upside joint bear all vertical loads. Shi C [38] carried out checking calculation and static simulation of finite element model for a new form of steel frame-precast concrete facade panels joint connection. The amplification coefficient of horizontal seismic action was set at 5.0. The bearing capacity of bolts, welded joints under tensile force, shear force and bending moment were checked. The research verified the available mechanical performance of this new form.

At present, the design of flexible connection is mostly in the form of elliptic bolt hole, which is convenient for deformation and energy dissipation of precast concrete facade panels. Xueqin W [39] proposed a new type of connection, in which 4 trusses of one-story and one-span assembled steel frame-precast concrete facade panel models were tested by pseudo-static test to verify the good deformation performance of the joint connection. Under the horizontal load, the concrete at the bolt joint was crushed and the expansion of bolt hole leaded to different displacement between the precast concrete facade panels and the main structure. The research showed that the joint connection was an important link for precast concrete facade panel to adapt to the deformation of steel frame. Jianyao G [29] suggested strengthening the connection between precast concrete facade panels and the main structure. Ailin Z [40] put forward a new type of assembled form suitable for the precast concrete facade panel. ABAQUS finite element model was established for the steel frame with precast concrete facade panels. The hysteretic curve was obtained by static and pseudo-static finite element analysis, and the feasibility of the new assembly was evaluated by ductility and energy dissipation capacity.

The existing structure design of connections is easy to cause stress concentration, and the slender bolt of joint connections will be the first failure area under the earthquake action, which reduces the bearing capacity of precast concrete facade panels.

4. Influence on seismic behaviour and deformation following ability of steel frame

This chapter mainly discusses the seismic performance of steel frame structure with precast concrete facade panels, and the deformation following ability of precast concrete facade panels while steel frame deforming, which are carried out according to shaking table test, pseudo-static test and finite element simulation.

4.1. Shaking table test

Guoqiang L [34-35] conducted shaking table tests on two two-story spatial steel frame-precast concrete facade panels models respectively in order to study the dynamic characteristics of precast concrete composite facade panels and precast ALC facade panels and the stress characteristics of flexible joint connections. The research showed that the higher the floor was, the greater the deformation was, and the inter-story displacement of the steel frame under the 0.6 g El-Centro earthquake was 1/54. Qiaoling X [41] carried out a shaking table test on a two-story full-size steel frame-precast concrete composite facade panel and found that after connecting the precast concrete facade panel, the natural vibration frequency decreased significantly because the increase of structure gravity. As the inter-story displacement of the steel frame-precast concrete facade panels was smaller than that of the steel frame, it was considered that the precast concrete facade panel had a good following deformation performance, but the following performance had little relationship with the tightening moment of the connection bolts. Yuliang Z [42] conducted a reduced scale shaking table test on a spatial steel frame-precast concrete facade panels model, and found that the increased precast concrete facade panels significantly reduced the amplification coefficient generated by the structure delivering from the opposite side seismic waves, so that it was obvious that precast concrete facade panels increased the lateral stiffness of the steel frame.

4.2. Pseudo-static test

The research objects include precast ALC wall panel, precast concrete composite facade panel and precast recycled concrete facade panel.

Guoqiang L [43] carried out static and pseudo-static tests on three trusses of steel frames and six trusses of two-story single-span steel frames with precast ALC panels, in order to study the influence of precast ALC panels and installing forms on the stress deformation and lateral stiffness of the main structure. The top angle on both ends of prefabricated ALC panels started to fall off while the load increasing. The results showed that the embedded ALC panel was superior to the external ALC panel in twice stiffness and improving 76.9% bearing capacity of the structure. The external ALC panel owned a better deformation ability.

Hetao H [44-45] carried out pseudo-static tests on 8 trusses of one-story one -span precast concrete composite facade panel steel frame model respectively. The specimens were all destroyed by the fracture of welding seam at the connections, and the spalling of precast concrete composite facade panel starting from the connections. He studied the influence of energy-saving composite panels and joint connections on the reciprocating performance of the main structure, evaluated the seismic performance of steel structure, and gave the recommended value range for equivalent damping coefficient in different stress circumstances. In order to study the seismic performance of the structure, Jian Z [46] conducted a pseudo-static test on the plane model of 4 trusses of one-story and one-span steel frame-precast composite facade panels, and found that the cross-bracing steel plate support of precast composite facade panels can improve seismic performance of the main structure. The hysteretic performance of precast concrete composite XPS panel with cross-bracing steel plate supporting the steel structure system was studied. The cross-steel plate is used to maintain stability of the precast concrete composite facade panel. According to the hysteretic pinching phenomenon of the specimens, it was found that the shear slip at the connection between the wall panel and the steel frame could slow down the structural damage. By comparing the stiffness degradation curve of pure steel frame, steel frame with composite panels, steel frame with composite panels and cross-bracing steel plate and the energy dissipation capacity of the structure, it was found that the composite panels with cross-bracing steel plate system improved the ultimate displacement of the whole structure and has better hysteresis performance. The independent slip deformation of the precast concrete composite facade panels in the post-loading period of the pseudo-static test reduced the damage of the steel frame, and the lateral stiffness increased by 1.2-2.9 times.

Hongchao G [25] conducted horizontal reciprocating test with axial compression on 3 trusses of one-story one-span scale model respectively, and obtained the effect of precast recycled concrete facade panel on the improvement of structural strength. During the test, the inter-story displacement, shear deformation at the top of the precast recycled concrete facade panels, the section and joint strain of structural components and the deformation at the joint connections of structural components were measured. The influence of precast recycled concrete facade panels and beam-column joints on the seismic performance of steel frame structures were studied. The seismic performance of the steel frame was evaluated according to its horizontal bearing capacity, lateral stiffness, displacement ductility and energy dissipation capacity. The steel frame with precast concrete facade panels had a 17% increasing in peak load capacity and a 16.7% increasing in displacement ductility and seismic performance at the later loading period.

4.3. Finite element simulation

Finite element model includes spatial model and flat model. SAP2000 finite element software is mainly used for finite element simulation of spatial steel frame-precast concrete facade panels model. ABAQUS and ANSYS finite element software is used for finite element simulation of flat model of flat steel frame-precast concrete facade panels and the details of the flexible joint connections.

In order to set up the finite element model of spatial steel frame-precast concrete facade panels, Yuliang Z [42] calculated SAP2000 for a two-story spatial model. The research showed that the top floor of spatial steel frame with precast concrete facade panels reduced the acceleration response of the top floor by about 16%, and the precast concrete facade panels produce lateral stiffness, precast concrete facade panels was beneficial to the seismic resistance of the main structure. In order to study the seismic performance of precast concrete facade panels and joint connections, Jiupeng L [47] conducted pseudo-static simulation of a two-story and two-span flat steel frame by ANAYS finite element software, and set up ground motion simulation of the spatial steel frame by using SAP2000. The inter-story displacement and the displacement response of the top floor under different working conditions were extracted.

Dianshen Z [48] carried out a numerical analysis using ANSYS finite element software based on the pseudo-static test of embedded and external precast ALC panels in reference [43]. The friction coefficient, which was more suitable for aerated concrete, was used in the simulation to build the aerated concrete constitutive model with descending section. The joint connections were presented as orthogonal duplexing spring in the simulation model. The finite element model had a high degree of fitting with the test results, and can reflect the internal crack development of ALC panels when it was damaged. The hysteretic curves showed that the pure steel frame was at elastic stage, and the steel frame with precast concrete composite facade panel had obvious ductility and yield point. Minglei W [49] used ABAQUS for numerical simulation of a single-story and single-span flat model. Shell elements were used to build the model, and full hysteretic curves were obtained. The stiffness degradation was 85.7% and the energy dissipation coefficient is 0.4. If the hysteretic curves of pure steel frame and steel frame with precast concrete composite facade panels could be compared, the energy dissipation comparison would be more ideal. Lijian S [50-51] carried out finite element simulation of pseudo-static test including axial load on the flat model of precast recycled concrete facade panel made of full-scaled steel frame with one-story and one-span on the basis of the test in reference [25]. The study doubted that the model did not show the crack and aggregate friction of precast recycled concrete facade panels accurately, so that the simulation results amplified the energy dissipation capacity of precast recycled concrete facade panels on the main structure.

At present, the influence of precast concrete facade panels on the seismic performance of the steel frame is mainly evaluated by the hysteretic curve, skeleton curve and other parameters obtained through experiments. In the test process, the failure of precast concrete facade panels and joint connections is the mark of test finishing. However, the deformation and bearing capacity of the connection area between the structural components and the precast concrete facade panels are not analysed in detail. The seismic response of the whole structure cannot be detected by the pseudo-static loading of the flat steel frame structure. The seismic performance evaluation should consider the acceleration response of the model.

5. Conclusion

This paper introduces the mechanical characteristics of assembled steel frame-precast concrete facade panels and summarizes the related research and exploration of Chinese scholars in recent years. The conclusions and suggestions are as follows:

(1) At present, the shaking table test of assembled steel frame-precast concrete facade panels model cannot be carried out until its failure, and the pseudo-static test cannot more accurately reflect the actual seismic performance. From one-span assembled steel frame-precast concrete facade panels flat model for pseudo-dynamic test, it is easy to compare with the main structure and displacement response of the acceleration response, as well as the deformation and bearing capacity differences of structures and precast concrete facade panels connection area. Pseudo-dynamic test could be better for precast concrete facade panels seismic performance research.

(2) The joints of flexible connection are bolted, which is easily lead to stress concentration. Design optimization should be carried out for bolt joints of flexible connections, such as thickening plates, reducing bending and avoiding irregular areas. The flexible connection can adapt to the deformation coordination of steel frame and reduce the degree of weakening of the rigidity of the precast concrete facade panels.

(3) The finite element simulation of assembled steel frame- precast concrete facade panels focuses on the analysis of SAP2000 as a spatial model and the analysis of ABAQUS and ANSYS as a flat model. However, there is no unified classification on the constitutive relation of precast concrete facade panels with different structures and the interaction relation of different forms of connectors. Assembled structure building for different situation, will produce different materials, different shapes of panels. How to find a more suitable form of precast concrete facade panel, how to find the common characteristics from the personality, will be the majority problem to solve in the future.

6. References

[1] Han J, Lihua C, Zuochao L and Yu L 2018 *Structural System and Practical Case of Assembly Buildings* (Nanjing: Southeast University Press) pp 93-105

[2] Jie W, Ning L, Han J, Yi Z, Dunjun W, Xuefei W, Liujin W, Leqi C and Jian Z 2018 *The Design and Application of Prefabricated Concrete Building* (Nanjing: Southeast University Press) pp 26-29
[3] Emad F G, Adrian M C, Colin F D and Graeme S 1999 *J. Struc. Eng* **125** 32-39
[4] Xiangdong T, Jerome F Hajjar, Arturo E S and Carol K S 2005 *J. Constructional Steel Research* **61** 531-552
[5] Guohua S, Ruoquan H, Gu Qiang, and Youzhen F 2011 *J. Const. Steel Research* **67** 1821-1834
[6] Zheng L, Minjuan H, Xijun W and Minghao L 2018 *J. Const. Steel Research* **140** 62-73
[7] Lei W, Zhenyun Tang, Yue L and Kai Q 2019 *Construction and Building Materials* **211** 756-770
[8] Takuya N, Hideo F, Kunio F and Shirou K 2010 Performance of an exterior metal curtain wall in seismic responses of a high-rise building *J. Technology & Design* **16** 535-540
[9] Zhenbao L, Leiling J, Lida T and Dongyan W 2017 Study on the structural properties of lightweight microporous concrete wall panels *17th National Symposium on Modern Structural Engineering Conf. Industrial Construction* (Tianjin) pp 66-70
[10] Shengcai L, Jianjing J and Qingrong Y 2001An Inquiry into Calculation Approaches of Bending Strength about Composite Shear Wall Slab *Structural Engineers* **3** 35-40
[11] Hetao H, Xiaojing H, Guoqiang L and Yanming W 2009 Ultimate Load-Bearing Capacity of the Energy-Saving Composite Sandwich Panels *Journal of Building Materials* **12** 106-111
[12] Guowei Z, Boshan C, Xiao W, Qisong M and Hui W 2016 Research on the flexural behavior of autoclaved aerated concrete external panel *Building Structure* **43** 97-102
[13] Jiasen L, Chenggong Z, and Zhengzhen P 2013 Analytical Methods of Precast Outer-Wall Hanging Panels *Chinese & Overseas Architecture* **1** 105-107
[14] Yanbo L, Weiqi G, Xiaodun W, Qing M and Qiong H 2018 Experimental Research on Wind Resistance and Finite Element Analysis of Reinforced Concrete Composite Insulation Wall Plate *Journal of Tianjin University (Science and Technology)* **51** 95-102
[15] *Code for Design of Concrete Structures* GB50010-2010
[16] *Standard for Design of Steel Structures* CHN Standard GB50017-2017
[17] *Code for Seismic Design of Buildings* CHN Standard GB50011-2010
[18] *Technical Standard for Assembled Buildings with Concrete Structure* CHN Standard GB/T51231-2016
[19] *Technical Specification for Precast Concrete Structures* CHN Standard JGJ1-2014
[20] *Technical Standard for Application of Precast Concrete Facade Panels* CHN Standard JGJ/T485-2018
[21] Hetao H, Bosheng L, and Jinwen L 2007 Problems on Measures About Energy-saving Composite Wall of Steel Residence *Steel Construction* **22** 41-44
[22] Bosheng L, Jinwen L, and Hetao H 2007 Study on Energy-saving Composite Panel in Steel Residence Houses *Steel Construction* **22** 9-12
[23] Youdong C, Yuanrong J, Weiqi F, Yongchao L, Hetao H, Canxing Q and Yanming W 2017 Current Issues and Suggestions for Sandwich Composite Panels of Steel Structure Residence *Industrial Construction* **47** 52-57
[24] Jinhu W, Jiasen Lu, Zhengdong X and Zhenpeng Z 2014 Upper Load-bearing Joint Design Method of the Precast Concrete Wall Panels *Building Structure* **44** 47-51
[25] Hongchao G, Lijian S, Yunhe L, Jiping H and Zhenshan W 2017 *Journal of Building Structures* **38** 63-73
[26] Wei W, Yueshi C, Yiyi C and Hetao H 2019 Full-scale Shaking Table Test of a Floor-by-floor Assembled Steel Braced Frame with External PC Composite Wall Panel *Journal of Building Structures* **40** 88-96
[27] Guoqiang L, Mingji F, Yijing L and Ye L 2005 *China Civil Engineering Journal* **38** 27-32
[28] Mingji F, Jingfeng W and Guoqiang L 2013 Shaking table test of steel frame with ALC external wall panels *J. of Constructional Steel Research* **80** 278-286

[29] Jianyao G 2016 Seismic Analysis of Prefabricated Exterior Board of the Assembled Frame Structure *M.S. thesis, Dept. Architect. and Civil Eng.*, Southwest Jiaotong Univ. China
[30] Lijian S, Hongchao G, Yunhe L and Changwei J 2017 *Chinese Journal of Applied Mechanics* **34** 742-747
[31] Lijian S 2016 Study on Seismic Behavior of Flexible Steel Frame with Recycled Concrete External Wall *M.S. thesis, Dept. Structure Eng., Xian Univ. of Tech.* Xian China
[32] Jingfeng W and Beibei L 2017 Cyclic testing of square CFST frames with ALC panel or block walls *J. Constructional Steel Research* **130** 264-279
[33] Jingfeng W, Beibei L and Jinchao L Experimental and analytical investigation of semi-rigid CFST frames with external SCWPs *J. Constructional Steel Research* **128** 289-304
[34] Guoqiang L, Xin Z, Feifei S, Wenli G, Zunquan Y and Shiwen J 2003 *Earthquake Engineering and Engineering Vibration* **23** 64-70
[35] Mingji F, and Guoqiang L 2008 Experimental research on seismic behavior of ALC external wall panels and masonry wall *Building Structure* **38** 55-58
[36] Yong J, and Caiyuan C 2008 *Structures Units & Units Architecture* **4** 42-46
[37] Yong J, and Caiyuan C 2009 *Construction Wall Innovation & Building Energy-Saving* **4** 42-46
[38] Shi C, Ganping S, Kunhong L, Shenggang F and Hong G 2017 *Building Structure* **47** 46-52
[39] Xueqin Wang, Xuechun L, Lin M and Xiaojun Z 2017 Experimental Research on Seismic Behavior of Prefabricated ALC External Panels *Steel Construction* **32** 22-26
[40] Ailin Z, Lin M, Xuechun L and Zinqin J 2016 *J. of Building Structures* **37** *suppl. 1* 152-157
[41] Qiaoling X, Yuanhui L, Zhenxi H and Zhong Y 2010 *J. Earthquake Engineering and Engineering Vibration* **30** 72-79
[42] Yuliang Z 2009 Shaking Table Test and Numerical Simulation of Steel Frame Assembled House with Exterior Wall *M.S. thesis, Dept. Structure Eng., Harbin Ins. of Tech.* China
[43] Guoqiang L and Cheng Wang 2005 The Hysteretic Behaviour of Steel Frames with ALC Out-hung and In-filled Walls *Steel Construction* **20** 52-56
[44] Hetao H, Canxing Q, Guoqiang L, and Jingfeng W 2012 Cyclic Test on Steel Frames with Energy-saving Sandwich Composite Panels *Engineering Mechanics* **29** 177-184
[45] Hetao H, Jian Z, Haitao Z, Jingjing L, Haining L, Zhonglong L and Xiping W 2014 *Engineering Mechanics* **31** 85-91
[46] Jian Z, Haitao Zang, Jingjing L, Minglei W and Zhonglong L 2014 *ISSF* pp 481-488
[47] Jiupeng L 2009 Study on Vibration Damping Properties of External Wall Panels of Industrialized Housing *M.S. thesis, Dept. Structure Eng., Tianjin Univ.* China
[48] Diansheng Z, Liang C and Weiwei W 2010 Hysteretic Behavior Analysis of Steel Frame Structure with ALC Walls *Journal of Zhejiang University of Technology* **38** 448-452
[49] Minglei W, Hetao H, Yanfei S and Qian W 2012 Parametric Analysis on Hysteretic Behavior of Steel Frame Structure with Energy-saving Composite Wallboard *ISSF* pp 433-440
[50] Lijain S, Hongchao G, Yunhe L and Yanan L 2017 *Earthquake Engineering and Engineering Dynamics* **37** 67-76
[51] Lijian S, Hongchao G and Yunhe L 2017 *Construction and Building Materials* **157** 790-808

Acknowledgement
Financial support from the National Key Research and Development Program of China (Project No. 2016YFC0701500) is appreciated.

IOP Conf. Series: Materials Science and Engineering **690** (2019) 012004 doi:10.1088/1757-899X/690/1/012004

Assessment on the Effect of Polyethylene Terephthalate and Low-density Polyethylene Plastic Waste as an Additive in Bituminous Paving Mixes

F D Santos[1*], M A M Taguba[1,2] and A R Alzona[1,2]

[1] Department of Civil Engineering, National University, Sampaloc, Manila 1008 Philippines
[2] Department of Environmental and Sanitary Engineering, National University, Sampaloc, Manila 1008 Philippines

Email: fdsantos@national-u.edu.ph

Abstract. The continuous growth of the population also brings about an increase in traffic that traverse our roads. Due to this increase in the volume of traffic, pavements are subjected to more distress which can lead to its failure. Studies have been performed in order to address these issues on material properties. At present time, waste disposal has been a challenge for various industries. And of the wastes that are abundant in our surroundings, plastic is proving to be a major problem because it poses a major threat to the environment. As a possible solution to the growing problem of the disposal of plastic wastes as well as the constant pursuit to improve the properties of materials, recycling and incorporating plastic to construction materials has been explored in various studies. This study aims to assess the effects of Low-Density Polyethylene (LDPE) and Polyethylene Terephthalate (PET) as an additive to Hot Mix Asphalt (HMA) on the stability, flow and bulk specific gravity of the paving mix. A parametric study was conducted in order to investigate the effect of additive concentration (4%, 7% and 10% by weight of asphalt binder), particle size (2.36 mm, 4.57 mm, and 9.5 mm) and mixing temperature (145^0 C, 160 ^{0} C, and 175 ^{0} C) on the stability, flow and bulk specific gravity of Plastic Waste Bitumen Binder (PWBB). Results show that adding PET and LDPE as additive increases the stability by 36.82%, improves the flow by 22% and it also increases the bulk specific gravity by 2.36% compared to a traditional bituminous mixture.

1. Introduction

The Philippines has been noted through a series of studies for having one of the fastest urbanizations in Southeast Asia [1]. One of the characteristics of urbanization is the continuous increase in the population which incurs an increase also in the traffic volume that travel the streets. The combined effects of the increase in the rate of motorization and private vehicle use intensifies even more the congestion that is present in roads [2]. Due to the high volumes of traffic, pavements experience higher distresses. Some of these serious distresses are rutting, shoving, stripping, and fatigue cracking which may lead to the complete failure of the pavement [3]. Such distresses can also reduce the performance of asphalt pavements under the effect of heavy traffic loading, high temperatures, and water damages. The deformable component properties of bitumen define the good performance of a road pavement [4]. One important property of bitumen mixtures in the wearing course design is its ability to resist shoving and rutting under traffic therefore the stability of the paving mix should be high enough to handle traffic adequately.

Nowadays, disposal of different plastic wastes produced from different industries is a major problem. Plastic wastes, specifically in urban areas can stay in the environment for a long time that can severely ruin the environment unless it is properly disposed of [5]. Due to the problem brought about by plastic wastes, studies have been conducted in order to find a solution by means of recycling and incorporating them in materials to provide a means of waste diversion. These studies were conducted because of the good mechanical properties of plastic. Molten plastics provide good binding property and increases the value of the compressive and bending strength of flexible pavements when used as a binder [6].

This study aimed to assess the effects of Polyethylene Terephthalate (PET) and Low-Density Polyethylene (LDPE) as additive on the stability and flow together with the bulk specific gravity of the Plastic Waste Blended Bitumen (PWBB) mixture. The method used is the standard mixing method focusing on the significant elements of the study. This paper also aimed to evaluate the effect of varying amount of PET and LDPE (4%, 7% and 10%) by weight of asphalt binder, particle size (2.36 mm, 4.57 mm, and 9.5 mm) and mixing temperature (145^0 C, $160\ ^0$ C, and $175\ ^0$ C) on the stability, flow and bulk specific gravity of PWBB and compare it to a control mix.

2. Materials

The materials used to produce the samples such as asphalt, cement filler, and aggregate as well the as testing equipment is located at the Department of Public Works and Highways Asphalt Section Laboratory of the Bureau of Research Studies in Quezon City. The PET and LDPE used in the study were collected by the researchers that came from containers of different beverages and junk food that are sold from the local stores and convenient stores in Manila.

3. Experimental methodology

3.1. Synthesis of plastic waste bitumen binder

The researchers fabricated samples of asphalt mixture cylinder infused with PET and LDPE to magnify the effectiveness of the data gathering instruments. The materials used in this research were no less than the materials used in actual practice. The materials consist of asphalt binder, cement filler, coarse and fine aggregates, collected PET and LDPE. The gathered PET and LDPE waste are then washed and dried. The dried materials were then shredded and cut into three different sizes from large (coarse), medium (fine) and smallest (finest), the shredded samples were then sieved through 3/8', Sieve No. 4 and Sieve No. 8 in order to categorize their particles size diameters.

3.2. Methodology on parametric samples

The concentration of PET and LDPE additive was varied from 4%, 7% and 10% by weight of asphalt binder with the ratio of distribution of 50% for both plastic types (PET and LDPE). Also, the mixing temperature was varied from 145℃, 165℃ and 175℃ in determining the properties of the bituminous pavement. Moreover, aside from the concentration and the mixing temperature, the varying particle diameter size size (2.36 mm, 4.57 mm, and 9.5 mm) was also examined on its possible effect on the mechanical properties of the bitumen.

3.3. Preparation of test specimen

The aggregates and asphalt were heated in the oven, mixed in a bowl using a metal spatula. The mixing temperature of the PWBB was varied from $145\ ^0$ C, $165\ ^0$ C, to $175\ ^0$ C. The mixed samples were then placed into the moulding cylinders where bitumen is added compacted on each circular face. Each side received 75 blows by the compaction machine in accordance with the standard. The researchers tested the cylinders consisting of various proportions and additive mixtures as determined in the aim of this paper.

Aside from the bulk specific gravity of the specimen, the bituminous pavement also underwent the Marshall Stability test to examine the stability and flow of the mixture. One by one the specimens were loaded perpendicularly into the Marshall testing apparatus. A certain force in now applied by the machine on the perpendicular side of the up to such a point that the specimen stopped resisting. The data was obtained from the machine and was recorded for analysis.

3.4. Testing of specimens

The standards adopted are those mandated by the American Society for Testing and Materials. ASTM D 6927 (Standard Test Method for Marshall Stability and Flow of Bituminous Mixtures) was the principal standard used as a basis for evaluating the strength and density of the modified specimens.

4. Results and discussion

4.1. Parametric study of PWBB.

The results of the stability test of PWBB with varying temperature, plastic concentration and particle size are shown in Figure 1.

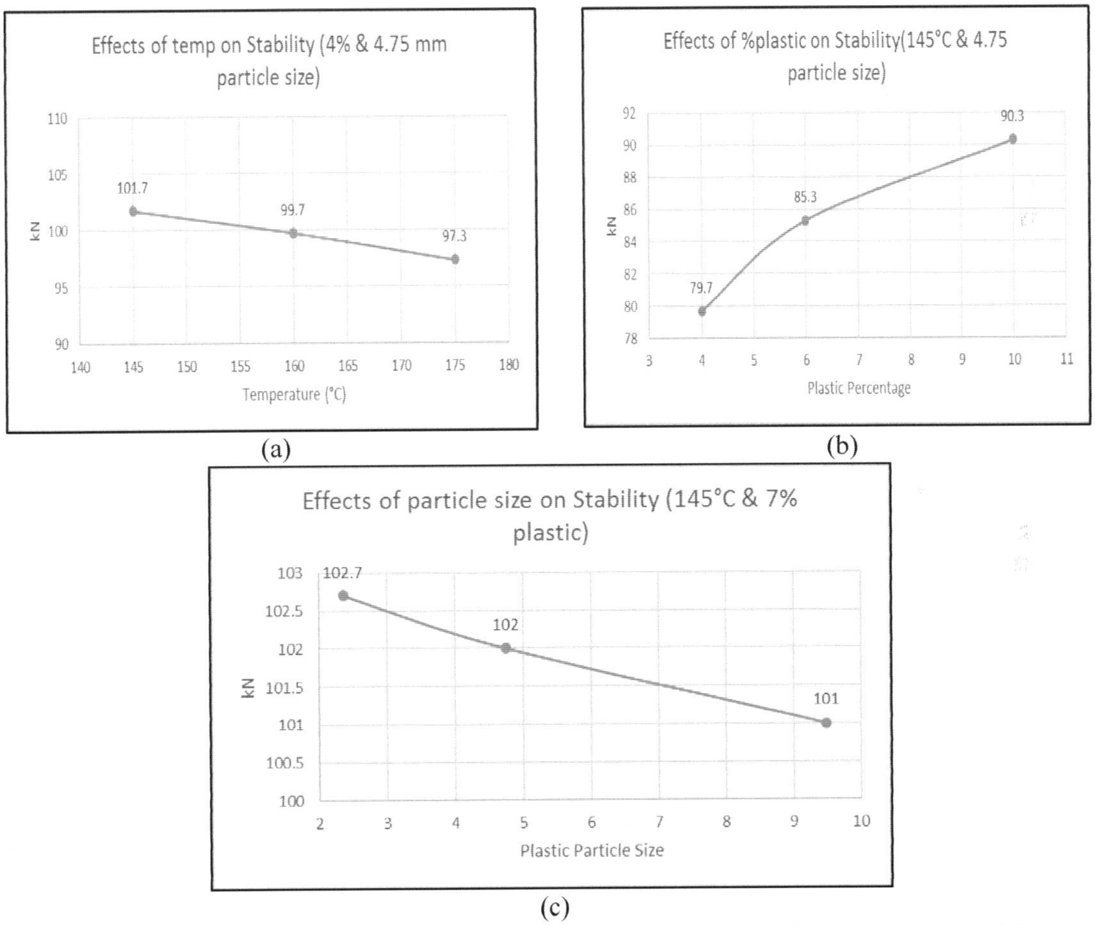

Figure 1. The effect of varying temperature, plastic concentration and particle size on stability

Based on the results, the effect of increasing mixing temperature and plastic particle size showed decreasing stability in PWBB, however, the while increasing the plastic percentage infused in the PWBB, the stability also increases up to a value of 90.3 kN. In comparison to the normal asphalt mix which recorded a stability of 66 kN, the stability of the PWBB increased by 36.82%. Figure 2 presents the effect of the varying parameters on the flow.

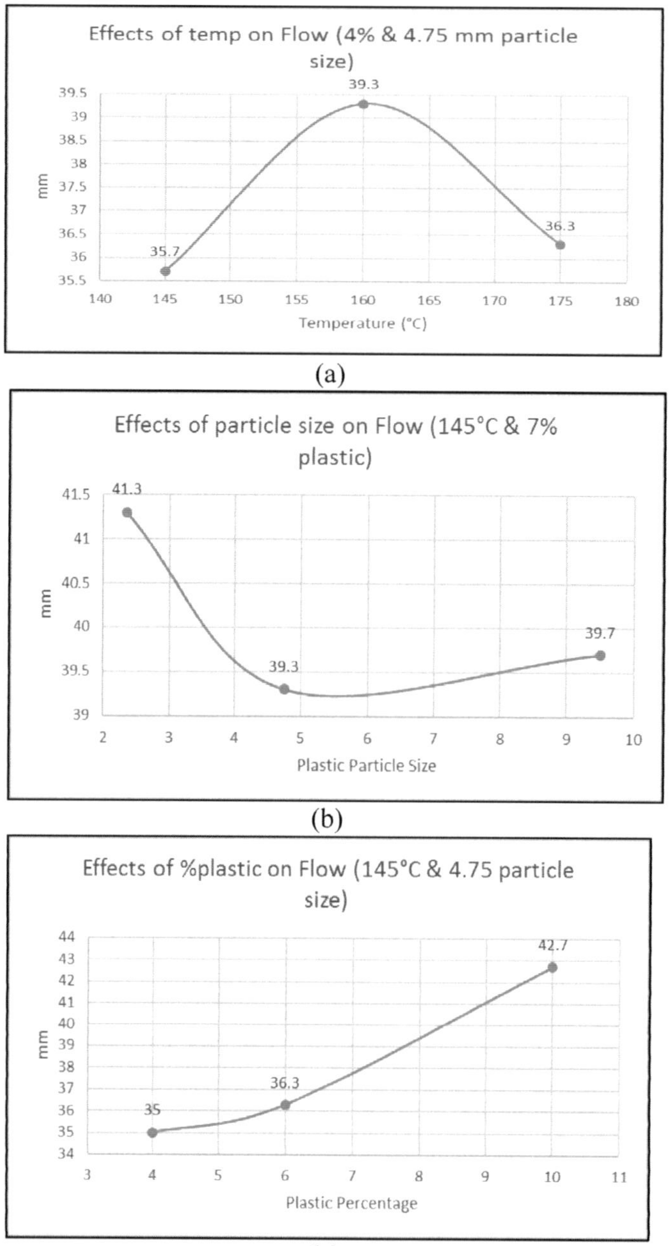

(a)

(b)

(c)

Figure 2. The effect of varying temperature, plastic concentration and particle size on flow

Results of the flow test show that increasing mixing temperature and plastic particle size again showed decreasing flow in PWBB, however, increasing the plastic percentage in the PWBB also increases the flow to a maximum of 42.7 mm. Comparing it to the flow of the control sample, the flow of PWBB is greater by 22%. A summary of the results of the Bulk Specific Gravity is depicted in Figure 3.

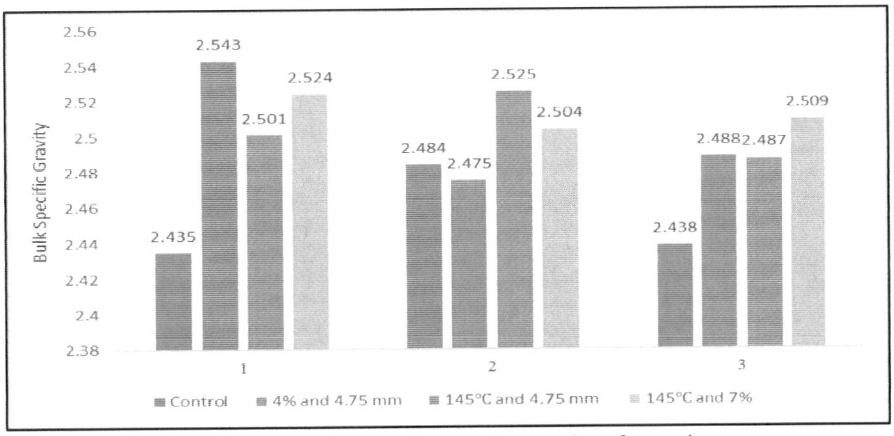

Figure 3. Bulk specific gravity results of samples

In the three test, results that the PWBB showed significant results even in Bulk Specific Gravity. Comparing the maximum bulk specific gravity obtained to the control sample, it can be said that it has increased by 2.36%.

5. Conclusion

This paper meant to evaluate the effect of using PET and LDPE as an additive in Hot Asphalt Mix on the mechanical properties such as stability, flow and bulk specific gravity and results were also compared to a traditional bituminous mixture. Considering the varying factors of temperature, plastic percentage and plastic particle size and results of the tests, the following conclusions were made:

- Due to the addition of PET and LDPE as additive in the bituminous mixture, there is an evident increase in the stability, flow and bulk specific gravity of HMA compared to normal (conventional) pavement mixtures.
- With respect to the concentration of plastic, increasing the amount of plastic in the mixture decreases the stability, flow and bulk specific gravity of the mixture. And Assessment on the effect of polyethylene terephthalate and low-density polyethylene plastic waste as an additive in bituminous paving mixes.
- Based on the parameter of particle size of plastic waste in the mixture, the increase in particle size is directly related to the increase in the stability, flow and bulk specific gravity of the paving mixture.

The findings also signify that plastic wastes can be used in the production of an innovative and low-cost construction material while serving as a waste diversion for plastic wastes in order to address the waste disposal problem of the community.

6. References

[1] Roberts B and Kanaley T 2006 Urbanization and Stability in Asia: Good Practice Approaches in Urban Region Development, *ADB*, Cities Alliance.

[2] Regidor J R F (2013) Traffic Congestion in Metro Manila: Is the UVVRP Still Effective? *Philippine Engineering Journal* **34**(1), 66-75.

[3] Kurikulum, D. (2012). Abstract, 14(1), 47–64. https://doi.org/10.1061/(ASCE) 0733-9410(1991)117.

[4] Kalantar, Z., Mahrez, A., & Karim, M. R. (2009). Properties of bituminous binder modified with waste polyethylene terephthalate (PET).

[5] Rajkumar, P. (2015). A Study of the Plastic Waste and Environmental Degradation. ABC *Journal of Advanced Research*, **4**(1), 9–16.

[6] Gawandea, A., Zamarea, G., Rengea, V.C., Saurabh Taydea, G. B. (2012). An overview of waste plastic utilization in asphalting of roads. *Journal of Engineering Research and Studies*, 3(2), 1–5.

IOP Conf. Series: Materials Science and Engineering **690** (2019) 012005 doi:10.1088/1757-899X/690/1/012005

Behaviour of the Undrained Shear Strength of Soft Clay Reinforced with Natural Fibre

N Suffri, M Jeludin and S Rahim

Civil Engineering Programme Area, Universiti Teknologi Brunei, Brunei Darussalam

Email: p20181002@student.utb.edu.bn

Abstract. The term soil reinforcement is conventional since decades ago for the stabilization of soft ground such as of clay and peat. Numerous research has arisen in the utilization of natural fibres as the reinforcement materials. Cost reduction, increment of sustainability awareness and eco-friendly environment are some of the advantages when using natural fibres to stabilize soft ground. A research study was carried out to evaluate the strength of the soft soil when unreinforced and reinforced using natural fibres. The findings on the experimental investigation of the study will be presented in this paper. Crushed coir fibres were used to reinforce an intermediate plasticity soft clay where both materials were collected locally in Brunei Darussalam. The crushed coir fibres were added at 0.5%, 1.0%, 1.5% and 2.0% to the dry weight of the sample. A series of an unconsolidated undrained Tri-axial test was conducted on the unreinforced and reinforced samples where the behaviour of the samples were observed and compared. The results indicated that inclusion of fibres affects the soil's undrained shear strength. It was observed that increasing the percentage inclusion increases the undrained shear strength of the soil, up to a certain amount. Further increment of fibres, however, does not show further improvement in the undrained shear strength.

1. Introduction

One of the challenges in construction on soft ground is that problems such as insufficient bearing capacity and soil settlement may arise. These complications are generated by the low shear strength and low permeability of the soil. The strength of the soil is a crucial factor in any construction fields in order to resist the high tension stresses imposed by the structures. A shear failure may occur along the internal surface of the soil, if its condition is not being enhanced, due to excessive compression.

Ground improvement technique, particularly soil reinforcement has been introduced to overcome these problems by improving the engineering properties of the soft ground to enhance the performance of the soil [1]. Reinforcement materials are added into the soil matrix to enhance the strength and decrease the soil's deformation [2]. This paper will present the study on soil reinforcement by randomly distributing the fibres. A random distribution of fibres also helps to develop the shear strength of the soil without having to present the weakness plane in the soil composite which eventually will help to reduce settlement [3]. Fibres that are naturally available from plants such as coir fibres, hemp, wheat and bamboo have gained attractions among researchers over the past few years due to the rise in sustainability awareness [2, 4–5]. Other reinforcement materials are of synthetic or manufactured fibres such as polyester, scrap rubber tire and polypropylene fibres but the production of these materials contribute to the emission of greenhouse gases. Hence, the option of using natural fibres is favoured [6].

Previous laboratory investigations have proven that the strength and stiffness of soft soil increased significantly with the addition of coir fibres [2, 7-8]. The use of hemp fibres added ductility to compacted soft clay by replacing the strain softening with strain hardening behaviour in the reinforced clay. The presence of hemp fibres improves the soil by increasing its fibres content from 0.5% to 1.25% but reduced at 1.5% showing the presence of threshold fibres content [9]. Another research was studied using sawdust as the fibres material where the unconfined compressive strength increases by 41.436%, from 0% to 3% addition of sawdust fibres into the clayey silt soil but when 5% of the fibres was added, the unconfined compressive strength decreased. A similar pattern was observed for the unconsolidated undrained shear strength but the percentage improvement from 0% to 3% was found to be 39.535% [10]. When the normal effective stress is higher, the effect of adding homopolymer polypropylene fibres becomes insignificant regardless the amount of added fibres. Based on the results from direct shear test, 0.25% addition of fibres is more effective than the 0.5% fibres [11]. Reinforcement using crushed coir fibres have been done on the assessment on the soft soil's consolidation behaviour where 2.0% inclusion of fibres gives the most significant effect [12]. This paper attempts to use a similar type of fibre reinforcement but on the investigation of the strength of soft clay using the unconsolidated undrained Tri-axial test.

2. Experimental programme

2.1. Materials used

Reinforcement fibres: Brown fibres from coconut husks or known as coir fibres were used in this study. The fibres were extracted from the outer shell of matured coconuts as they have stronger properties and high resilient. Since coconuts are in abundance in Brunei Darussalam, some of the unwanted husks were disposed by burning which pollutes the air, hence utilizing them as reinforcement materials helps to protect the environment and reduce cost as they are naturally available. However, the length-to-diameter ratio of the fibres vary as the thickness are not consistent. This gives problems to the observation of the experiment if the fibres were to be used in strands of different lengths. Hence, the coir fibres used for this study were crushed into tiny grains.

Soft soil: The testing sample for this experiment was natural clay which is slightly dark brown in color, obtained locally at the site of the Raja Isteri Pengiran Anak Saleha (RIPAS) Bridge at Sungai Kebun. The sample was first studied according to BS 1377: Part 2 to determine its basic properties as tabulated in Table 1.

Table 1. Properties of natural clay.

Properties	Values
Liquid limit, LL	38.4%
Plastic limit, PL	22.5%
Plasticity index, PI	15.9%
Optimum moisture content, w_{opt}	20.0%
Specific gravity, G_s	2.7
BSCS Classification	Intermediate plasticity

2.2. Materials preparation

Reinforcement fibres: Strands of coir fibres were extracted from matured coconut husks that have been washed, dried and decorticated. The fibres strands were further crushed and passed through a 600 μm sieve size to eliminate the remaining strands of fibres.

Soft soil: The wet soft clay was brought into the laboratory and oven-dried for 24 hours prior to crushing and sieving using the 63 μm mechanical sieve shaker. In accordance to BS 1377: Part 4, the clay sample was compacted using the standard Proctor test where the optimum moisture content was obtained. The dry clay sample was then mixed with water at its optimum amount and kept in a sealed bag for 24 hours to retain its moisture before preparing the testing sample.

2.3. Specimens preparation

Crushed coir fibres were randomly mixed with the wet soil sample after 24 hours. This is to imitate the in-situ state where the fibres can only be added into the ground in its wet condition. The mixing of the fibres with the soil was done thoroughly in portions to ensure that all of the fibres required were fully used in the required amount of soil sample in the Tri-axial mould having a diameter of 50 mm and 100 mm height. The samples were compacted in layers of 3 following the standard Proctor test and the testing specimens were extruded using the Tri-axial mould that has been greased.

2.4. Testing programme

The undrained shear strength of the unreinforced and reinforced specimens was investigated by conducting an unconsolidated undrained test using the Tri-axial equipment in accordance to BS EN ISO 17892 (Part 8) at a strain rate of 0.5mm/min. The crushed coir fibres were added at 0.5% increment to a maximum percentage inclusion of the fibres at 2.0% considering the previous work done on the consolidation behaviour of the unreinforced and reinforced specimens [12, 13]. A repeat test with 0.5% inclusion of fibres was also conducted in order to check for the results' consistency.

3. Test results

The results of the fibre-reinforced Sungai Kebun Clay specimens are presented in Table 2 where q_u is taken at 20% axial strain. From this table, the unconfined compressive strength is influenced by the amount of fibre content in the sample.

Table 2. Results of unconfined undrained test.

Test Number	Inclusion of fibres to dry weight of soil (%)	q_u (kPa) Taken at 20% axial strain
1	0.0	40.0
2	0.5	56.9
3	1.0	57.0
4	1.5	89.4
5	2.0	117.7

The typical stress-strain relationship of the fibre-reinforced specimens is shown in Fig.1. As seen in the figure, the amount of fibres included in the specimens significantly contribute to the increase in stress for all various percentage of fibre inclusions. The strength increases from 40.0 kPa when unreinforced to 56.9 kPa when 0.5% of crushed coir fibres was added, and increases further to 89.4 kPa at 1.5% and 117.7 kPa at 2% of crushed coir fibres inclusion. Thus, the stress-strain behavior of the fibre-reinforced specimens depends on the content of fibres. This result correlates well with [14] where their compressive strength increases for 0.4% to 1.6% inclusion. The compressive strength increases by 2 times for the inclusion of dry coir fibres and NaOH-treated coir fibres, and 3 times improvement for CCl_4-treated coir fibres indicating that the fibre inclusion influences the behaviour.

Fig. 2 shows the degree of improvement by inclusions of coir fibres. It can be represented by taking q_u/q_{uo}, where q_u is taken as the maximum compressive strength at 20% axial strain and q_{uo} is the maximum unconfined compressive strength with no fibre inclusions. As can be seen in the figure, the inclusion of fibre has an effect on the strength of the soil where increasing the fibre content improves its strength. The latter could be due to the effect of fibre strains has on the friction between the fibres and soil grains. The soil and fibres are bound closely together, minimizing the void content which makes shearing of soil difficult. The fibres are high in tensile strength such that when the specimen is sheared the fibres are able to withstand the tension and thus, improve the shearing strength of the specimen. Previous research [10], however, indicated that compressive strength would increase as amount of fibre inclusion increases up to a certain percentage, beyond this percentage there will not be any significant improvement. The amount of inclusions analyzed in this study is up to 2% only. As suggested, there is a limiting fibre content where beyond that value there will not be any improvement in strength and this limiting value is affected by the type of fibres being used. In order to understand the behavior further, the amount of inclusion in this study need to be further increased.

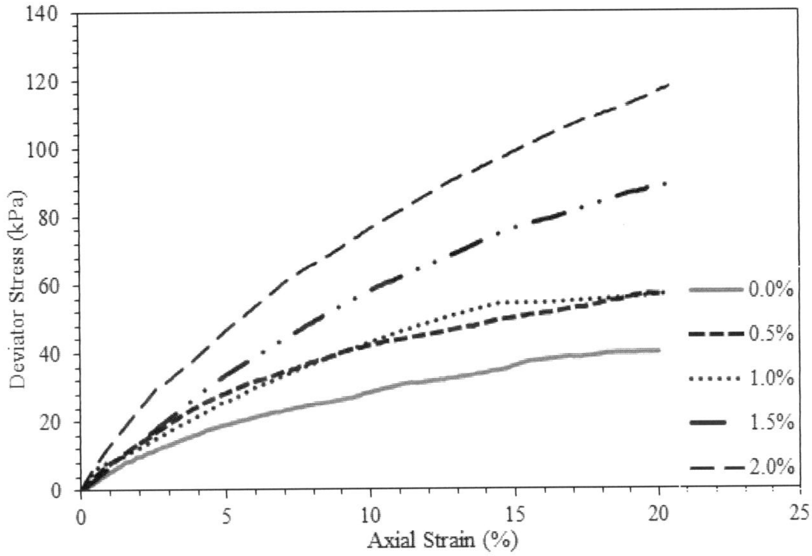

Figure 1. The stress-strain relationship of the unreinforced and fibre-reinforced specimens.

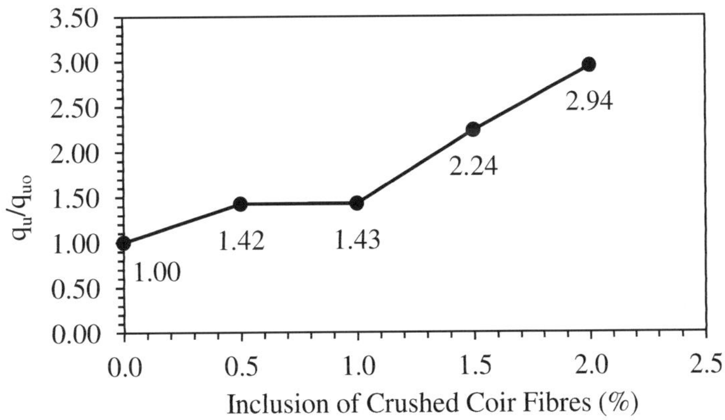

Figure 2. The relationship between percentage inclusion of fibres and the degree of improvement.

4. Conclusion

From the results that have been discussed, the following conclusions can be drawn:

- The utilization of crushed coir fibres gives effect on the behaviour of the unconsolidated undrained Tri-axial test where the addition of crushed coir fibres as the soil reinforcement shows a significant improvement on the strength behaviour when compared to being unreinforced.
- The unconfined compressive strength of the unreinforced specimen improved significantly when 0.5% of crushed coir fibres was added up to 2.0% and the axial strains increase with the increase in stresses for all percentage increments of crushed coir fibres inclusions.

5. Limitations and recommendations

- The existing displacement transducer limits the movement to take further readings of the tests conducted where the tests had to stop before reaching the actual failure stage yet. It is recommended to take into account that the existing transducer needs to be modified or replaced.
- Since the tests were conducted at 100 kPa chamber pressure for all testing specimens where only the percentages inclusion of fibres act as the variation, to obtain the cohesion additional chamber pressure should be applied at 200 kPa and 300 kPa. This enables further investigations on the behaviour samples.

6. References

[1] Vidal H 1969 *The Principle of Reinforced Earth, National Academy of Sciences* **282** 1-16
[2] Kar R K, Pradhan P K and Naik A 2012 *Consolidation Characteristics of Fibre Reinforced Cohesive Soil', Electronic Journal of Geotechnical Engineering* **17** 3861-3874
[3] Gowthaman S, Nakashima K and Kawasaki S 2018 *A State-of-the-Art Review on Soil Reinforcement Technology Using Natural Plant Fiber Materials: Past Findings, Present Trends and Future Directions, Materials* **11 (4)** 553
[4] Anggraini V, Asadi A, Huat B and Nahazanan H 2015 *Effects of Coir Fibers on Tensile and Compressive Strength of Lime Treated Soft Soil, Measurement* **59** 372-381
[5] Maurya S, Sharma D, Jain D and Kumar D 2015 *Review on Stabilization of Soil Using Coir Fiber, International Journal of Engineering Research* **4 (6)** 296-299
[6] Hejazi S, Sheikhzadeh M, Abtahi S and Zadhoush A 2012 *A simple review of soil reinforcement by using natural and synthetic fibres, Construction and Building Materials* **30** 100-116
[7] Nahta S and Vibhakar N 2014 *Study of the Effect of Coir Fibre Reinforcement on the Strength Parameters and CBR Value of Clayey Soil, International Journal of Engineering Research & Technology (IJERT)* **3 (4)**
[8] Sivakumar Babu G and Vasudevan A 2008 *Strength and Stiffness Response of Coir Fiber-Reinforced Tropical Soil, Journal of Materials in Civil Engineering* **20 (9)** 571-577
[9] Abou Diab A, Sadek S, Najjar S and Abou Daya M 2016 *Undrained shear strength characteristics of compacted clay reinforced with natural hemp fibers, International Journal of Geotechnical Engineering* **10(3)** 263-270
[10] Jasim O and Çetin D 2016 *Effect of Sawdust Usage on the Shear Strength Behavior of Clayey Silt Soil, Sigma Journal Engineering and Natural Sciences* **34 (1)** 31-41
[11] Mirzababaei M, Arulrajah A, Haque A, Nimbalkar S and Mohajerani A 2018 *Effect of fiber reinforcement on shear strength and void ratio of soft clay, Geosynthetics International* **25 (4)** 471-480
[12] Suffri N, Jeludin M and Rahim S 2018 *Consolidation behaviour of soft clay reinforced with crushed coir fibres, 7th Brunei International Conference on Engineering and Technology*
[13] Jeludin M, Suffri N and Rahim S 2019 *The Consolidation Properties of Natural Fibre Clay Composite, Proceedings of the 4th World Congress on Civil, Structural, and Environmental Engineering*
[14] Rakesh Kumar D, Vishwas Nandkishor K and Gayathri V 2012 *Effect of Addition of Treated Coir Fibres on the Compression Behaviour of Clay, Jordan Journal of Civil Engineering* **6 (4)** 476-488

Acknowledgment
The authors wish to thank the Civil Engineering Programme Area at the Faculty of Engineering at Universiti Teknologi Brunei for providing research facilities and for supporting this research study.

IOP Conf. Series: Materials Science and Engineering **690** (2019) 012006 doi:10.1088/1757-899X/690/1/012006

Effect of Street Canyon Configurations and Orientations on Urban Wind Velocity in Bangkok Suburb Areas

D Jareemit[1] and M Srivanit[2]

[1,2] Faculty of Architecture and Planning, Thammasat University, Pathumthani, Thailand.

Email: jdaranee@gmail.com

Abstract. Urban ventilation is considered as one parameter, which impacts on building energy consumption and outdoor living condition. This study primarily investigates the effect of street canyon characteristics as well as its orientations on wind pattern and velocity in street canyon via using urban microclimate simulation model, ENVI-met. The wind simulations are performed with four aspect ratios or height-to-width ratio (H/W), three canyon lengths, and four canyon orientations. Calculated wind velocity measured at 1.5 m height of the centre of the street and pedestrian on both sides of the street ranges from 0.2 to 0.97 m/s. N-S and NW-SE canyons, which oriented parallel to the prevailing wind has greater wind velocity than those oriented perpendicular to the wind. The wind velocity is sensitive to aspect ratio and canyon length. The wind velocity in shallow canyon is mostly higher than those in the deep canyon. However, increasing the canyon length considerably improves the low wind speed in the deep canyon up to twice.

1. Introduction

Bangkok and suburb provinces is affected by urban heat effect and traffic congestion results in air pollution problem. Breathable city is a concept to bring more ventilation in the city. Pedestrian wind velocity is importance that the wind can remove heat and also dilute pollutant concentration from the traffic, which those make better environment for outdoor living. Previous studies have shown that urban configurations have a significant effect on airflow patterns and wind speed in street canyon [1,2,3]. High density cities typically have poor air ventilation since the buildings obstruct the wind flows. Lower site coverage ratio could increase the pedestrian ventilation performance [1]. In addition, wider street canyon also provides better ventilation potential than those in shorter one [2,3]. Rahman et al. [4] and Kitous and Adolphe [5] investigated the effect of aspect ratio of the length to width (L/W) on wind velocity and found that the short canyon contributed to low wind speeds when compared to that of the long canyon. Their findings agree well with Kitous et al.'s results [6].

The study of the effects of canyon configurations on wind velocity have been widely studied in several countries. Base on the literature reviews on design achieving urban ventilation, most focused on aspect ratios of building height to street width in relation to canyon orientations. One study considered building length to street width. Such the design combinations of street width, building height and its length have not been investigated yet. In addition, previous investigations were conducted with limitations by local land use and regulations, which such conditions might differ and not properly implement in Thailand's context.

The aim of this study firstly investigates an effect of street canyon configurations regarding a proportion of street width, building height and its length on pedestrian wind velocity. Lastly, the urban wind velocities influenced by different urban typologies are compared and discussed. The results of this finding will be extended for further evaluation the design configurations that promote cooling and thermal conditions for outdoor living. Regarding the case studies used in this investigation, the urban typologies are constructed based on local regulation and those forms are typically found in Bangkok suburb areas. This research results could benefit for urban planners and architects to determine characteristics of buildings and planning, which can create Bangkok suburb areas a breathable city.

2. Research methodology

The street canyon is usually expressed by aspect ratio (ratio of building height to street width, H/W) and the length (L) of the canyon. In this study, urban wind simulations of forty-eight street canyon characteristics classified by four aspect ratios of 0.5, 0.7, 0.9, and 1.1, three canyon lengths represented by building block size of 4, 5, and 10 units per block, and four canyon orientations of N-S, NS-EW, E-W, and NW-SE (shown in figure 1) are performed using urban microclimate simulation model, ENVI-met model. Those classifications of urban forms and its geometries generally represent the typical characteristics of residential cluster developments in Bangkok neighbourhood areas.

The ENVI-met model [7] has been accepted in evaluating thermal environment and outdoor thermal comfort. In addition, this simulation model is widely used for a measure of airflow and wind velocity in urban scale. Average 10-years local meteorological data for the Rangsit region obtained from a nearest weather station of the Thai Meteorological Department (TMD) at Pathumthani Station (14.116°N, 100.621°E), such as wind direction, air temperature, wind velocity, long and short wave radiation is used as an input in ENVI-met modelling. From the simulation results, the wind speeds are measured across the street canyon at 1.5 m. height of both pedestrian paths and centre of the street as presented in figure 2. The simulation results are validated by comparing field measurements. The correlation between measured data and simulation shows a strongly correlation with $R^2 = 0.91$ ($p <$ 0.001).

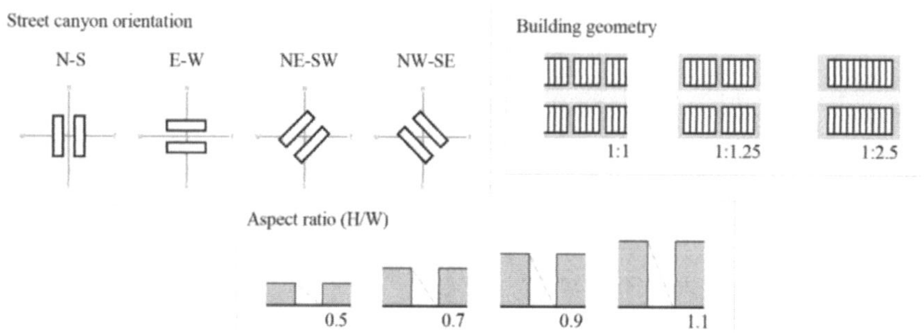

Figure 1. Design parameters and input variables used to perform in ENVI-met simulation.

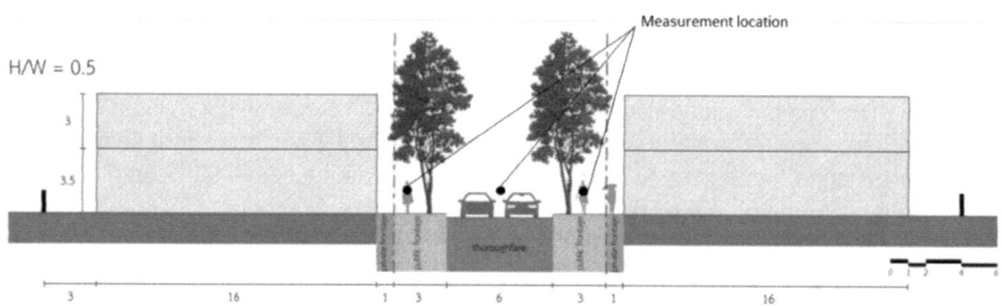

Figure 2. Section of street canyon and measurement locations.

3. Results and discussion

Wind simulations through ENVI-met modelling are used for examining and comparing level of wind speeds in street canyon for different street orientations, aspect ratios, and the canyon lengths. The results are presented and discussed as the three following topics.

3.1. Effect of street canyon orientations on wind velocities

Figure 3 (left) presents distribution of colour shades visualizing wind speeds in street canyon at 1.5 m. height of the 1.1 canyon with 4-units length as an example. Average wind velocity along forty-eight canyon typologies is calculated. It is found that wind velocities are typically high at the front of canyon and the velocities gradually reduce at the end of the canyon. The street canyon orientations parallel to local wind mainly from the south, specifically the canyon oriented N-S and NW-SE, have the highest wind velocities (ranging from 0.7-1 m/s) while the NE-SW and E-W canyons which oriented perpendicular to the wind have the lowest velocities, which ranges from 0.2-0.4 m/s.

Considering airflow pattern in these four canyon orientations, the wind mostly flows narrow to the street since characteristic of the street acts as a tunnel. However, at street intersection, the airflow does not deflect the direction to perpendicular streets since the proportion of street width to the canyon length might be insufficient to allow the wind changes that direction [8]. For all street canyon arrangements, the distribution of wind velocities across the canyon vertically increases with building height while the velocities in the horizontal plane are almost the same. The wind velocities near the ground level are lowest, which are less than 0.58 m/s and the wind speeds at the rooftop are above 1.00 m/s.

Figure 4 (right) compares the box plots of the wind velocities of four different aspect ratios classified by four different street canyon orientations. The wind velocity in the street canyon is sensitive to the canyon aspect ratio. The distribution range of wind velocities in aspect ratio of 1.1 is wider than others (standard deviation = 0.05) while the wind velocities in the canyon with aspect ratio of 0.5 has small distribution range with standard deviation = 0.1. The distribution of wind velocities in N-S and NW-SE oriented canyons is larger than those of NE-SW and E-W canyons since orienting canyon parallel to the local wind could allow more wind to access into the canyon.

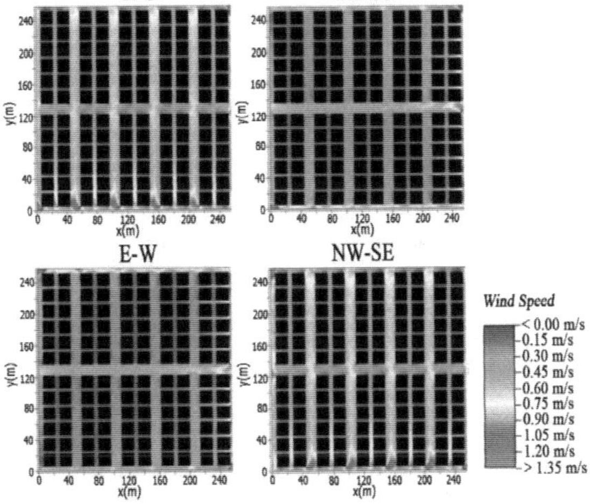

Figure 3. Calculation of cooling energy consumption associated to building geometries and street canyon orientations.

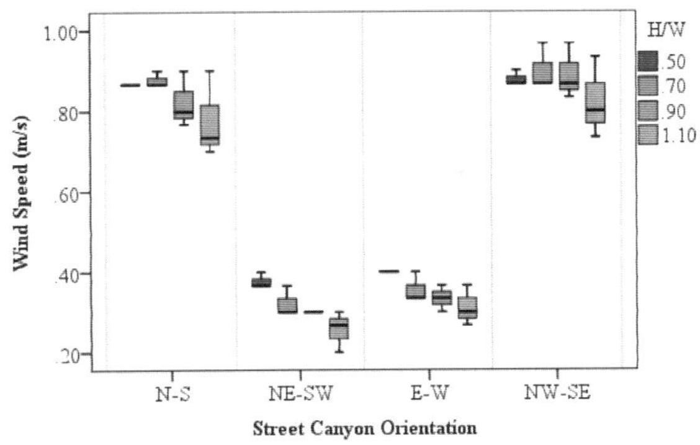

Figure 4. Box plot of the distribution of wind velocities in different street canyon orientations and aspect ratios.

3.2. Effect of canyon aspect ratios and canyon lengths on wind velocities

The study compares calculated wind velocities at 1.5 m height for four different aspect ratios of 0.5, 0.7, 0.9, and 1.1 with three canyon lengths of 4, 5, and 10 units per block. As mentioned above, the wind velocity profile across the canyon section increases in line with the building height. Overall, design of 10-units block length provides the highest pedestrian wind velocities to both deep and shallow canyons ranging from 0.9-1 m/s for N-S and NW-SE orientations and 0.3-0.4 m/s for E-W and NE-SW orientations (shown in figure 5). However, the wind velocities in the shallow canyons (H/W = 0.5 and 0.7) are slightly lower than those of the deep canyons. Such increment has similar results as found in Kitous et al. [6] and Rahman et al. [4] that street length significantly influenced more airflow velocity than in those of the shorter one.

Regarding wind performance in 4 and 5 units block lengths, the wind velocities at pedestrian level in the aspect ratio of 1.1 are lowest when compared to those of the shallow canyons. The pedestrian wind velocities in the canyon with H/W=1.1 range from 0.2-0.3 m/s for E-W and NE-SW orientations and 0.7-0.8 m/s for N-S and NW-SE orientations. Decreasing the canyon height significantly improves the velocity performance in E-W and NE-SW orientations up to 45 percent and 19 percent for N-S and NW-SE oriented canyons. High aspect ratios with long canyon length considerably increases the wind velocity than those of lower aspect ratios with the short length since the canyon shape might act like the small opening channel. This small hole accelerates the wind speed as called Venturi effect, which is related to proportion of canyon's width, length, and its height [8].

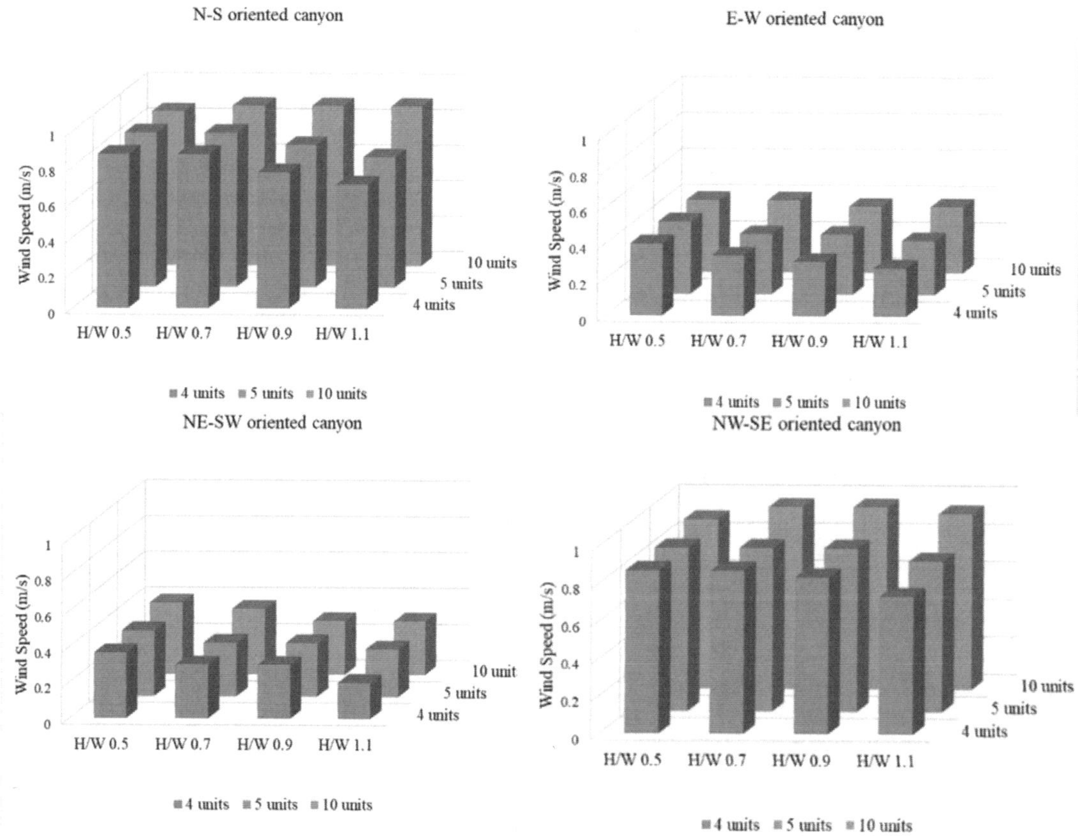

Figure 5. Calculation of wind velocities associated to street canyon length and aspect ratios distribution by canyon orientations.

3.3. Design strategies for urban ventilation

Increasing ventilation performance in pedestrian level for the street oriented parallel to prevailing wind, design the canyon with either 10-units length with H/W =1.1 and 4 units length with H/W=0.5 is recommended since it allows better wind to access into the street canyon than other design scenarios. The wind velocities in the canyon with 4-units length with H/W=0.5 are slightly lower. In contrast, for the street situated perpendicular to the wind direction, long canyon length (10 units per block) with H/W = 0.5 provides the highest wind velocities; however, the pedestrian wind velocities are much lower than those in the canyons parallel to the wind. To improve the wind speed in the street oriented perpendicular to the wind direction, other techniques such as providing more open space, step the building back, or increasing more opening between the canyon block might increase the wind velocity. The influence of such design combinations should be investigated in future study.

4. Conclusion

The study assesses wind velocities in street canyon, which influenced by various urban design configurations typically found in Bangkok suburb areas. The urban microclimate simulation model, ENVI-met is used to perform wind velocity as well as airflow pattern. The design of aspect ratio associated to canyon length considerably impact on the pedestrian wind. Most shallow canyons present high wind velocities than those of the deep canyons, which the maximum velocity for those parallel and perpendicular to the wind is 1.1 m/s and 0.38 m/s, respectively. The street canyon with aspect ratio of 1.1 and length of 4 units per block has the lowest velocity performance. Increasing the canyon length to 10 units per block effectively improves the wind velocity in the canyon by twice. Alternative design techniques for improving ventilation potential in street canyon oriented perpendicular to the wind should be investigated for further work. The limitation of this study is to assess wind velocity in canyon, which has the same building height. Such urban form is typically found in suburb areas, but building height in the big city varies. Consequently, future studies should account the influence of different building height, which might contribute different wind velocities and airflow patterns in street canyon.

5. References

[1] C Yuan and E Ng 2012 Building porosity for better urban ventilation in high-density cities – A computational parametric study *Building and Environment* **50** 176–189

[2] N Shishegar 2013 Street Design and Urban Microclimate: Analyzing the Effects of Street Geometry and Orientation on Air flow and Solar Access in Urban Canyons *Journal of Clean Energy Technologies* 52–56

[3] J A Akubue 2019 Effects of Street Geometry on Airflow Regimes for Natural Ventilation in Three Different Street Configurations in Enugu City *Different Strategies of Housing Design*

[4] W B Rahman, A Dinapradipta and I Defiana 2018 The impact of long canyon design and waterfront barrier buildings on wind velocity in a coastal area of Selatpanhang city *ARSITEKTURA* **6(2)**, 231–238

[5] S Kitous, R Bensalem and L Adolphe 2012 Airflow patterns within a complex urban topography under hot and dry climate in the Algerian Sahara *Building and Environment* **56** 162–175

[6] C A Arkon and Ü Özkol 2014 Effect of urban geometry on pedestrian-level wind velocity *Architectural Science Review* **57(1)** 4–19

[7] M Bruse 1999 The influences of local environmental design on microclimate- development of a prognostic numerical Model ENVI-met for the simulation of Wind, temperature and humidity distribution in urban structures. *Ph.D. thesis*, University of Bochum, Germany (in German)

[8] A W Spirn 1986 *Air quality at street-level: Strategies for Urban Design Boston*, Prepared for: Boston Redevelopment Authority

Acknowledgment

The authors wish to thank Pichamon Leethongin. This work was supported in part by a grant from the National Research Council of Thailand under the NRCT 2014 Grant No. P-15-50076.

IOP Conf. Series: Materials Science and Engineering **690** (2019) 012007 doi:10.1088/1757-899X/690/1/012007

Traditional Settlement Morphology from the Perspective of Thermodynamic Architecture Theory: Taking Two Villages in the Iberian Peninsula as Examples

Meiting He[1]

[1]College of architecture and urban planning, Tongji University, Shanghai, 200092, China

Email: hmt@tongji.edu.cn

Abstract. The form of rural settlements is not determined by a single factor, but by various factors. With a particular focus on settlement morphology and natural energy, this study introduces thermodynamic architecture theory to discuss the relationship between rural settlement morphology and environmental energy, starting from the analysis of the relationship between rural settlement morphology and building density, building order and settlement boundary dispersion, trying to provide a reference for the updated design of traditional settlements. The first part of this paper explains the research method of settlement form based on thermodynamic architecture theory from the theoretical level trying to construct the analysis framework of settlement form typology based on energy concept. The second part introduces the coupling between thermodynamic energy concept and morphology from the methodological level. The relationship between the three energy correlation factors of intensity, order and dispersion and settlement morphology is studied. In this paper, two settlement cases are quantitatively analysed by simulation and mathematical analysis. In conclusion, it proposes a new perspective for the morphological analysis of rural settlements, which provides a feasible framework for further research on energy utilization and microclimate comfort of settlements. This study also provides a preliminary concept for the integration of the buildings type and thermodynamic study of vernacular architecture in the future.

1. Introduction

The creation of an ideal environment is more embodied in a specific spatial organization (Amos Rapoport, 1969). For the general traditional vernacular settlement, many formations are through individual behaviour, that is to say, neither through government planning nor through the design of professional architects. In the long process of human history, traditional settlements have obviously formed their own unique patterns and forms.

The form of vernacular settlements is not determined by a single factor, but by various factors, including social factors, technical conditions, geography and climate factors. This paper argues that the combination of factors will vary in different climatic environments. Even Rapoport's theory, whose theoretical framework is very similar to Semper's mathematical formula, mentioned that climate is by no means a decisive factor. Semper proposed Y = F (x, y, z, etc), in which Y represents works of art, which is determined by the constant F and the variable (x, y, z, etc). Semper regards that the constant represents functions, and the functions are differentiated according to types. The variables are materials, places, nationalities, climate, beliefs and politics, etc. As for Amos Rapoport, he regards social culture as the constant, or "the primary factor", and other material and technological factors as variables, or "the secondary factor". However, the natural forces that human beings growing up in different regions need to overcome are different, and the demand for shelter is different as well. Therefore, "climate adaptation measure" can naturally serve as an effective concept to satisfy the demand. If this concept is expressed in diagrams, the degree of freedom of choice of the settlement form will be in the climate perspective, spanning from intolerable climatic conditions to the climatic conditions without necessity of consideration. For example, the weather condition will affect the form diversity of residential buildings to some extent as Figure 1 below, so the form selection will form a certain linear relationship [1]. Of course, in Rapoport's framework, each form of solution is based on specific social needs and technological resources, for example, the locals' location of settlements is generally based on personal cultural awareness and needs.

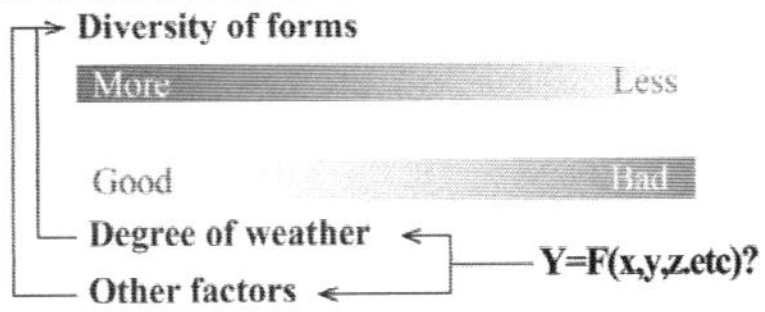

Figure 1. Diversity of Settlement Form.

This paper argues that if the potential logic and correlation between the form of vernacular settlements and climate adaptation can be explored, the influence factors of the form of a settlement can be dismantled to some extent, whether directly or indirectly. Therefore, the form of vernacular settlements is varied and mysterious based on different selection factors. It is often difficult to analyse its internal causes, and its generation generally follows the mode of selection, type, imitation and improvement over time.

2. Settlement morphology research method

This paper here proposes a new framework of traditional settlement morphology research from the perspective of thermodynamic architecture theory based on many predecessors' study of vernacular architecture and environment. There are many ways to study the influence and cause of settlements' forms and architectural forms based on climate diversity and thermodynamic theory. Firstly, based on different climatic types, we can investigate them such as hot-dry climate, hot-humid climate, continental climate, temperate climate, tropical climate, etc. and discuss the typical solutions of each region in terms of the necessary conditions of settlement morphology, spatial types and materials, which is a common method. In addition, we can discuss the geographical location and climatic conditions of various settlement forms in different climatic grades, and examine how people treat the climatic variables that lead to different climatic types, including temperature, humidity, wind, rain, radiation intensity and light. When the discussion enters micro level of natural energy, we need to introduce the concept of thermodynamic architecture to analyse the object. Because only by considering the settlements and buildings studied as an open interface system and carefully analysing the energy exchange between them and the external environment, can we understand the local settlements as a whole.

2.1. Thermodynamic architecture theory and vernacular settlement

Thermodynamics is an important part of physics. The etymology of thermodynamics is complicated. In 1849, the adjective part of the conjunction symbol appeared for the first time- thermo-dynamic. Then in 1854-1859, it was composed of its noun form thermo-dynamics to describe the subject of heat and energy movement, and hence the name Thermodynamics. Thermodynamics studies the relationship between heat, temperature, energy and work, the equilibrium and quasi-equilibrium states of matter, and the physical and chemical reactions in the interaction between the system and the outside world when changes occur [2]. It reveals the conservation relationship of energy conversion and points out the direction, conditions and limits of energy conversion process. It is of great significance to use thermodynamics to analyse and guide human creative activities [3]. When the concept of thermodynamics is introduced into architecture, it is called Thermodynamics Architecture. Thermodynamics architecture is a branch of human thermodynamics. Its research object is

IOP Conf. Series: Materials Science and Engineering **690** (2019) 012007 doi:10.1088/1757-899X/690/1/012007

architectural design theory, energy, entropy and thermodynamic law in single building and urban planning design. Alan Wilson published the first edition of Entropy in Urban and Regional Modelling in 1970, applying the method of entropy maximization in statistics to the study of urban and regional models. And describes the theory of thermodynamics architecture. He believes that the whole system theory and entropy have a positive effect on urban planning [4].

The research objects of thermodynamics architecture include naturale and climate elements, architectural form and material, human body perception, etc. In this paper, three thermodynamics factors, light, wind and heat (humidity), are proposed to be introduced into the concern of vernacular settlements. "Thermodynamics Intervention in the Study of Vernacular Architectural Form" is proposed, which defines the intersection of thermodynamics, climate architecture and vernacular settlements at the level of settlement scale. This paper attempts to combine theory, technology and analysis to answer the question below: What is the significance of natural energy intervention in vernacular architecture? How do the natural energy elements interact with the form of vernacular architecture and settlement? Is there some hidden logic between them?

2.2. Research and mathematical analysis of vernacular settlement form

In this part, a framework of settlement morphology analysis based on thermodynamic energy elements will be proposed, which explains how the three energy elements of light, wind and heat interact with three indicators: density, order and interface dispersion. The framework of analysis is shown in Figure 2.

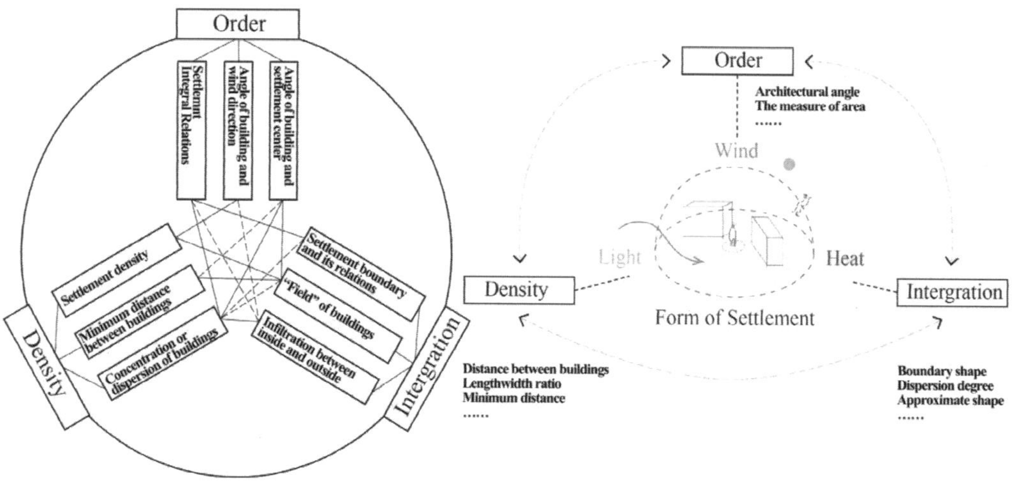

Figure 2. Settlement Form Relationship with energy elements and Three Indicators.

2.2.1. Light and settlement density. Light, which is conducive to keeping human health and creating a comfortable environment, is also one of the important climatic factors in the long history. In term of architecture, local people grasp the characteristics of light to ensure that it can be optimized in function and aesthetics, or to appropriately reduce and control excessive light in accordance with local cultural and functional needs. In term of settlement, the space and alley formed between buildings have a great influence on the light. Therefore, the minimum average distance between buildings in the settlement is taken as one of the indicators to evaluate the density of the settlement.

2.2.2. Wind and settlement order. Wind is needed by people, but sometimes it brings disaster to human beings. The relationship between settlement form and wind has been studied by many scholars, which is considered that there is a great correlation. When it's in hot summer, cool breezes are welcomed, but in cold winter, the wind makes people feel even more bitter cold. Building orientation and architecture design can reduce the negative impact of wind to a certain extent, and can further enhance its positive impact by promoting ventilation and maintaining the stability of the building structure. Therefore, the order degree of settlement is mainly studied from the perspective of architecture.

IOP Conf. Series: Materials Science and Engineering **690** (2019) 012007 doi:10.1088/1757-899X/690/1/012007

There are sometimes many criteria for judging the order of building settlements. If each building is oriented in the same direction, then obviously such a settlement has a high sense of order. But a circular settlement around the centre of a settlement, even though the orientation of each building is different from that of the northward direction, their centripetal magnitude is similar, so we can call such a settlement highly ordered. Therefore, when we judge the order degree of the settlement by the angle, we need to consider the average angle of each building and local dominant wind direction and the average centripetal quantity of each building and settlement center at the same time. Admittedly, if the settlement has multiple centres, such as the old village of Cuenca in Spain, as we will see later, then each building should be calculated with the nearest center.

2.2.3. Hot-wet and settlement dispersion. Rural architecture based on the concept of bioclimatology has been developed and used by many civilizations in the world for centuries [5]. Thermodynamic "heat balance" often refers to the dynamic balance of heat in a space, which has become a hot topic in the world. The balance between heat supply and emission is often reflected in the planning scale and the building settlement scale, and is also of great significance in the more micro scale of human body and building scale [6-8]. Just as Inaki Abalos puts forward the "Bodily doctrine" of thermodynamics in 2015, people's judgment of comfort is both physical and sensory [9].

The theory of dissipative structure was put forward by Prigogine in his article Structure, dissipation and life in 1969 [10]. The internal organization of the thermodynamic open system of entropy flow direction is very important. When this thermodynamic viewpoint is applied to the design of buildings and settlements, the internal organization of the dissipative structure corresponds to the organization mode of energy flow, which is a generalized form of buildings and settlements. At this time, we need to first investigate the boundary state of the system. Therefore, this paper calculates the shape of the settlement by computer, and analyses the discrete degree of the ellipse which is similar to it, so as to analyze the boundary shape of the settlement.

3. Analysis and result

There have been many related studies on settlement morphology around the world, mainly from the perspective of mathematical and geometric analysis of settlement morphology and building typology [11]. This part puts forward the three indicators mentioned above and focuses on light, wind and heat (wet) in the thermodynamic vernacular settlements, and introduces them into the morphological analysis of local settlements, taking two rural settlements as examples which can be seen in figure 3. The two cases selected in this paper are the old town of Cuenca in central Spain and the old town of Gavieira in northern Portugal, both in Iberian Peninsula. They have similar latitudes and geographical conditions, but there are obvious differences in climatic conditions. For example, in terms of the annual average predominant wind direction, Portugal's case is northward, while Spain's is eastward. In addition, the Gavieira settlement is located in the low-lying valley, while the Cuenca settlement in Spain is located in the valley between the two rivers.

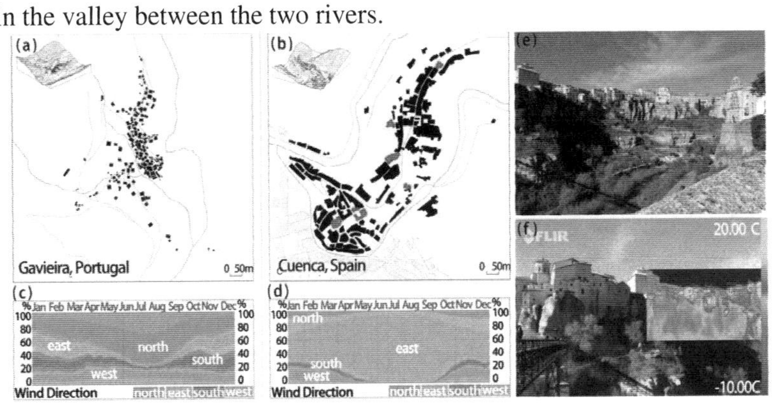

Figure 3. (a) Layout of Gavieira, Portugal, (b) Layout of Cuenca, Spain, (c) Wind Direction Map of Gavieira, Portugal, (d) Wind Direction Map of Cuenca, Spain, (f) Photo of Cuenca, (e) Thermal imaging of Cuenca

The following is the process of calculating the density index of settlement form. Firstly, based on Rhino-Grasshopper computer simulation technology, the boundary lines of each building unit are picked up, and the shortest distance between them is calculated, each of them has its own shortest distance. Secondly, because the distance between each building is numerically calculated, if there are n buildings in a settlement, the shortest distance per building will be n-1. Obviously, there is no basis for considering all the shortest distances in terms of density, which will lead to the inclusion of distances between two buildings that are far apart. Therefore, only the shortest distance of the first two times is considered here. Thirdly, by taking the average and standard deviation of all distances calculated and comparing the data obtained from different settlements, we can get the method of comparing the density of settlements. According to the above calculation, the average distance between each building of Cuenca settlement in Spain and Gavieira settlement in Portugal is 14.896 m and 9.241 m, respectively. The shortest distance between the settlements is 0.745m and 0.376m, the standard deviation is 30.10 and 17.36, and the dispersion coefficient is 2 and 1.88. That is to say, the density of Cuenca settlement in Spain is not only lower than that of Gavieira, but also the density difference of its settlements is relatively large.

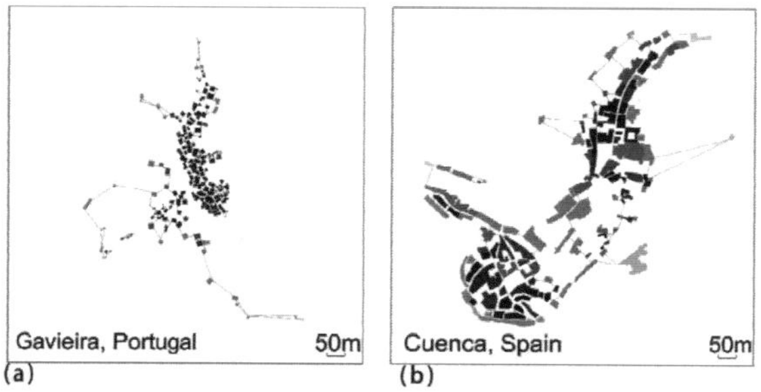

Figure 4. (a) Visualization of the Shortest Distance and Density in Gavieira, Portugal (b) Visualization of the Shortest Distance and Density in Cuenca, Spain

Figure 4 above is a density visualization of Gavieira and Cuenca settlements. The buildings with smaller average distances are shown in dark while the buildings with larger average distances are shown in light. The building with the darker colours can be understood as being denser in the entire settlement.

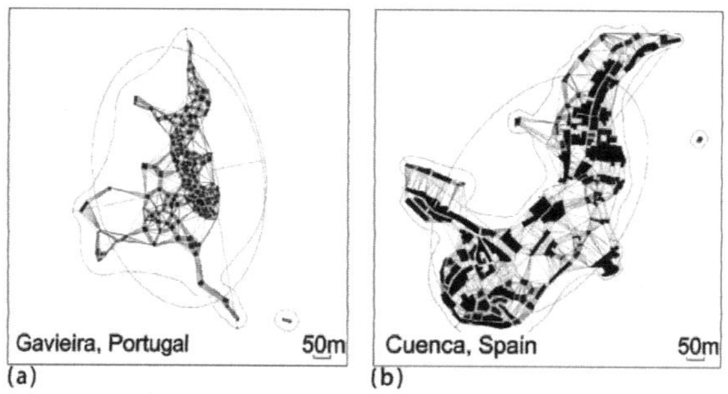

Figure 5. (a) dispersion diagram of Gavieira village in Portugal (b) dispersion diagram of Cuenca village in Spain

This paper describes how to calculate the dispersion index of settlement form, which can also be understood as boundary integration index. In order to further consider the overall energy utilization and comfort of the settlement in future research, the boundary of the settlement needs to be defined. Generally speaking, rural settlements are greatly influenced by natural factors as above. The shape boundary of natural rural settlements is often difficult to identify unlike many artificially controlled boundary forms in urban development. Firstly, this paper uses the outline of each single building extracted before to find the outline of the settlement interface. It is assumed that each building has an attraction to the periphery so the two buildings will form the sense of place between them. Therefore, the definition of boundary is not only the hard boundary of the building, but also the "field" adjacent to each building. Finally, morphological discretization analysis is carried out between the obtained settlement boundary and the approximate ellipse with the same aspect ratio and the same area, and the shape index and the degree of boundary discretization are obtained. Here, the shape index reflects the degree of morphological deviation between the settlement boundary and the similar ellipse with the same area. The minimum value of the shape index is 1. The closer the value is, the closer the figure is to the similar ellipse shape. According to the above calculation, the shape index of Cuenca settlement in Spain and Gavieira settlement in Portugal are 1.7 and 1.2 respectively, and the dispersion degree is 0.41 and 0.16. Figure 5 illustrates the dispersion of Gavieira and Cuenca villages. Therefore, through the calculation of the above methods, this paper finds that the boundary of Cuenca settlement is more concave and convex than Gavieira, and is more complex. The external boundary of Cuenca settlement can not only permeate with the internal buildings of the settlement, but also contact with the external environment more thoroughly.

4. Conclusion

Traditional cities and settlements can often be seen to deliberately adjust density to cope with climate. The traditional "cell-breeding" architecture and the "inward" high-density layout of small atriums in North African cities are designed to cope with the hot and dry desert climate and to withstand the hot summer wind, winter cold wind and dust accompanying the wind. Similarly, in order to obtain ventilation to cope with hot and humid climate, Dai villages in China specially make single buildings ventilated on all sides to achieve a "scattered" low-density layout.

In this paper, the thermodynamic architecture theory is introduced into the settlement morphology layout, and three methods are proposed to analyse the settlement morphology: light corresponding to density, wind corresponding to order and heat corresponding to dispersion. Firstly, the thermodynamic indexes of settlement morphology can be quantified at the level of mathematical and physical analysis. Through calculation, the density of Cuenca settlement in Spain is not only lower than that of Gavieira, but also the change of building distance is relatively large. Secondly, through the quantitative analysis of settlement morphology, the settlement morphology can be further qualitative and classified. With the help of the concept of thermodynamic open system, the calculated settlement boundary can describe the permeability of the boundary with the internal building and external environment to a certain extent, and thus can reflect the degree of its influence by external forces. Therefore, the correlation of these three indicators is helpful for the detailed description and classification of settlement morphology. It provides a feasible framework and basis for further study of settlement energy utilization and microclimate comfort.

5. Reference

[1] Rapoport A 1969 *House form and Cultua* (New Delhi: Prentice-Hall of India Private Ltd) p 84
[2] LI Linxue 2015 Knowledge, Discourse and Paradigm: Historical Perspective and Contemporary Frontier of Energy and Thermodynamics Architecture *Times Architecture* **2** 10-16
[3] Chen Zhicong 1998 On Thermodynamics in Civil Architecture *Engineering Physics* **8** 4
[4] Wilson A 2011 vol 1 *Entropy in urban and regional modelling* (London: Routledge)
[5] Singh M K, Mahapatra S and Atreya S K 2009 Bioclimatism and vernacular architecture of north-east India *Building and Environment* **44(5)** 878-888
[6] Kristensen E T, Møller W F, Rotne G, Sørensen P, Albjerg N, Mathiasen N K and Dahl T 2010 *Climate and architecture* (London: Routledge)
[7] Fathy H 1986 *Natural energy and vernacular architecture* ed Walter S and Sultan A A (Chicago: The University of Chicago Press)
[8] Nicol J F and Humphreys M A 2002 Adaptive thermal comfort and sustainable thermal standards for buildings *Energy and buildings* **34(6)** 563-572
[9] Abalos I and Snetkiewicz R 2015 *Abalos+Sentkiewicz: Essays on Thermodynamics, Architecture and Beauty* (New York: Actar Publishers)
[10] Prigogine I 1969 Structure, dissipation and life *Theoretical physics and biology* **23** 52
[11] Zhou J, Lin J, Cui S, Qiu Q and Zhao Q 2013 Exploring the relationship between urban transportation energy consumption and transition of settlement morphology: A case study on Xiamen Island, China *Habitat international* **37** 70-79.

Jakarta Groundwater Basin Recharge – Discharge Boundary Area Map: A Preliminary Study

Horas Yosua[1], Herr Soeryantono[2] and Dwinanti Rika Marthanty[2]

[1]University of Indonesia, Master of engineering student in civil engineering, Faculty of engineering, Depok, West Java, Indonesia
[2]University of Indonesia, Faculty of engineering, Depok, West Java, Indonesia

Corresponding author Email: horas_josua_sitorus@yahoo.co.id

Abstract. The most important component in groundwater basin are recharge and discharge area. The direction of recharge component is downward while discharge is upward. To discover local recharge and discharge boundary area of the Jakarta groundwater basin, this experimental research is using many cross sections and some hydraulic parameters of many studies that has been conducted before, then simulated it in finite element program SEEP2D (GMS) and finally validated it with bore log and monitoring wells. Ten cross sections selected from 3 studies based on head contour, stratigraphy and groundwater river interaction to be simulated in finite element method SEEP2D (GMS) then the result is validated. Since the result of the investigation is not finished yet, this paper wants to demonstrate the methodology that being conducted in the research with the horizontal map hypothesis. The map hypothesis is evident compare to the real situation of Jakarta groundwater basin today and it is approved that the method that is being used in this research is eligible and appropriate to find the boundary area of recharge and discharge. In the future this recharge and discharge boundary area horizontal map of Jakarta groundwater basin result will encourage many practical solutions in supporting Jakarta groundwater basin management.

1. Introduction

A groundwater basin consists of recharge area and discharge area. In a recharge area there is a component in the direction of groundwater flow near the surface that is downward, otherwise in a discharge area there is a component in the direction of groundwater flow near the surface that is upward [1]. Topography, geology or combination between them determine the character and location of recharge and discharge area [1].

The study of Jakarta groundwater within a 1439 km^2 area has been observed gradually since more than 2 decades ago. The outcome of the investigation produces delineation of regional recharge and discharge area of Jakarta groundwater basin in the example [2], [3] and [4] but the recharge discharge boundary area map of the Jakarta groundwater basin have not been invented yet until today. To discover local recharge and discharge boundary area in the Jakarta groundwater basin, this experimental research is using many cross sections and some hydraulic parameters of many studies that has been conducted before, then simulated it in finite element program SEEP2D (GMS) and finally validated it with bore log and monitoring wells. Since this recharge discharge boundary area map of the Jakarta groundwater basin has not existed yet in any studies before, this developing research is fundamental to develop a better groundwater management in the Jakarta groundwater basin.

2. Theoretical consideration and related researches

2.1. Groundwater flow model for Jakarta groundwater basin
The flow equations for all aquifer types are obtained from two basic principle continuity and Darcy's law. While continuity demands the conservation of mass, Darcy's law states that in an isotropic porous medium the specific flow rate is proportional to the negative head gradient [1].

2.1.1. Two Dimensional unsteady, saturated, confined, isotropic flow for horizontal direction. The equation of flow for unsteady through a saturated anisotropic porous medium with homogeneous and isotropic soil characteristic [1] is :

$$\frac{\partial^2 h}{\partial x^2} + \frac{\partial^2 h}{\partial y^2} = \frac{S}{T}\frac{\partial h}{\partial t} \quad (1)$$

where h = Piezometric head, S = Storage coefficient, and T = Transmissivity
Analytical solution for unsteady, saturated, confined, homogenous is known as the Theis solution :

$$h_0 - h\,(r,t) = \frac{Q}{4\pi T} W(u) \quad (2)$$

where W = Well function, T = Transmissivity, Q = Pumping rate

2.1.2. Two Dimensional unsteady, saturated, unconfined, isotrophic flow for vertical direction. Groundwater general equation for unsteady, unconfined, aquifer [1] is mathematically describe as:

$$\frac{\partial}{\partial x}\left(\frac{\partial h}{\partial x}\right) + \frac{\partial}{\partial y}\left(\frac{\partial h}{\partial y}\right) = \frac{n_e}{Kh}\frac{\partial h}{\partial t} \quad (3)$$

where n_e is effective porosoty, h is pressure head, K is hydraulic conductivity and x,y is Cartesian coordinate. The equation known as Boussinesq's approximation equation [5]. Because this is a nonlinear equation, it is not possible to obtain an analytical solution in general, but Neuman (1972) approached the general equation above with the analytical solution:

$$h_0 - h = \frac{Q}{4\pi T} W(u_a, u_b, u_n) \quad (4)$$

where $W(u_a, u_b, u_n)$ = Unconfined function, T = Transmissivity, Q = Pumping rate

2.2. Development of data set for simulator needs of Jakarta Groundwater Basin Management (2018)
Riko Apriatresnayanto in his thesis [2] conducted groundwater research on the Jakarta CAT with Parameters input in Modflow simulation for run test Jakarta groundwater basin simulator consist of boundary condition both horizontal and vertical boundary, land elevation (contour), initial condition, general head, specific head, groundwater aquifer parameters such as hydraulic conductivity as well as specific storage / yield, sources and sink such as rivers, lakes, wells which are included of recharge and discharge.

In accordance with Darcy's principle that the direction of groundwater has a beginning at high groundwater pressure, the recharge area leads to the area with the lowest groundwater pressure, namely the discharge area. The model formulation and planning process is carried out by selecting cross section in area where there is an increase or decrease in groundwater head pressure. In figure 1, for example, the eastern part of the south has a higher water pressure than the surrounding area. Because the area is an area with the highest groundwater, the area is a recharge area. Conversely the area with the lowest groundwater pressure in Figure 1 is depicted in blue color, then the area is a discharge area. Five locations with either increase or decrease groundwater head pressure is formed into a cross section of each with south – north orientation, then the five cross sections will be simulated in SEEP2D (GMS).

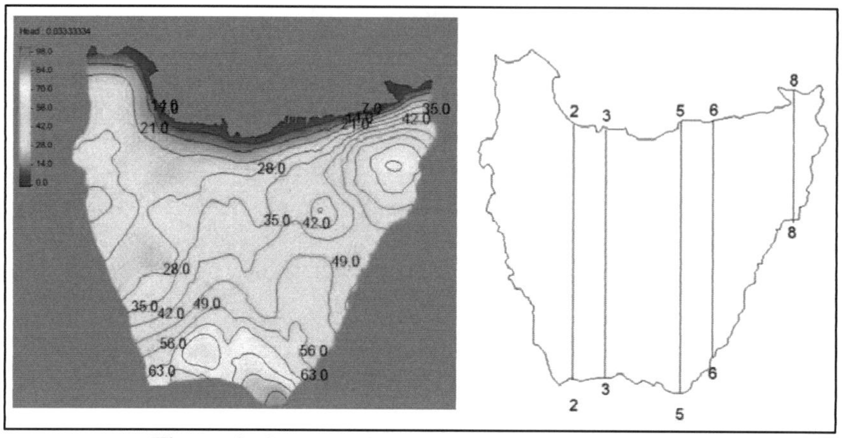

Figure 1. Cross section based on simulator (2018)

2.3. Stratigraphy and Hydro stratigraphy Groundwater Jakarta (2002)

In the results of research by Lambok hutasoit et al [3] shows a stratigraphic sequence from old to young in the Jakarta groundwater basin. Based on the dominance of lithology and stratigraphic position of each rock unit found rock units which are aquifers, axles and bedrock units.

Refer to lithology composition and stratigraphy position, unit rock that in charge of aquifer in Jakarta Groundwater basin are Kaliwangu middle Formation, Genteng Formation, Serpong Formation, Citalang Formation, and Endapan Volcanic Quarter. Unit rock that role as an akitard are Kaliwangu upper and lower Formation. Unit rock that role as a bed rock are Bojongmanik Formation, Jatiluhur Formation, Parigi Formation, Subang Formation, and Basal Gunung Dago as locally. Stratigraphy and Hydro stratigraphy Groundwater research founded 4 zones in Jakarta groundwater basin. First Zone (Aquifer 1) include Citalang Formation and Quaternary Volcanic Deposits. Second zone (Akuitard 1) is Upper Kaliwangu Formation. Third zone (Aquifer 2) consisted of Middle Kaliwangu Formation and Serpong Formation. The lowest zone, which is the fourth zone (Akuitard 2) is Lower Kaliwangu Formation.

All 3 south – north cross sections resulted in Stratigraphy and Hydro stratigraphy of the Jakarta Groundwater basin research [5] are being calculated numerically with SEEP2D (GMS) and also will be validated using data bore log and monitoring wells.

2.4. Groundwater – surface water interactions of Ciliwung River streams, Bogor Jakarta (2013)

The methodology applied in that study [4] was a combination of surface field measurements consisting of geological observations, groundwater flow net analysis, and hydro chemical analysis. Field measurements were taken at 65 stations, consisting of groundwater samples from digging wells at left and right part of the river bank and river water samples. Observations were conducted in May to August 2006 and were selected based on geology, according to the regional geological map and groundwater well availability.

Based on groundwater flow net analysis, the team identifies three groundwater and river water relations at the Ciliwung River stream. First type (effluent stream) that was identified from Bogor to Katulampa (Segment 1) which has the groundwater moved out to the river. Second type (Perched stream) was recognized at the University of Indonesia (UI) area (Segment 2). Results from geoelectrical survey suggested that aquifer layers were not directly in contact with the river basement. It was located on average 5 m below the riverbed. The last type is the Third type (influent stream) that was divided into four sub segments: from UI to Pasar Minggu, Pasar Minggu to Salemba, Salemba to Mangga Besar, and Mangga Besar to Muara. All locations exposed influent stream type. This segment was generally covered by alluvium deposit, which played the role of aquifer. The aquifers were located directly under the river stream. Based on water level measurements, river water level was higher than groundwater (influent stream type). River water infiltrated the underlying aquifers in divergent pattern, with < 0.1 % of hydraulic gradient.

IOP Conf. Series: Materials Science and Engineering **690** (2019) 012008 doi:10.1088/1757-899X/690/1/012008

From 3 west – east direction in the interaction of groundwater and water surface ciliwung river, this research will use 2 of them which are perched stream and influent stream. Effluent stream is not being used in this research because of most of the area not within the territory of the Jakarta Groundwater basin. The 2 cross river ciliwung cross section segment are being simulated and will be validated in this research to find the recharge – discharge boundary area of the Jakarta groundwater basin.

3. Recharge – discharge cross sections map methodology and modelling

The formulation and planning of this model are planned by determining the cross section that will be modeled in the SEEP2D (GMS) program. There are ten (10) cross sections needed for determining recharge discharge boundary areas modeling, including: five (5) North – South cross sections based on the result of the simulator 2018 [4] which is cross sections 2 – 2, 3 – 3, 5 – 5 and 8 – 8; three (3) North – South cross sections based on the results of the stratigraphic and hydro stratigraphic research [4] which is cross section 1 – 1, 4 – 4 and 7 – 7; and the last two (2) West – East cross sections based on the result of groundwater flow and the ciliwung river interaction research [2] which includes cross section 9 – 9 and 10 – 10. All the cross sections are delineated in figure 2.

The stage of methodology of this research includes: (1) Review (2) Formulation and Model Planning (3) Initialing Model Software (4) Building Model (5) Calibration and (6) Validation. First step to the fourth step is called Preparation Step. Preparation step before modeling is to collect the values of the horizontal hydraulic conductivity (K 1) and the values of the vertical hydraulic conductivity (K 2) based on previous researches and literatures. Different values of K 1 and K 2 will be taken to developing of each model. For five cross sections from the simulator [4], the result from lithology correlation Hydrogeology Unit (HGU) and soil type in borehole module (MTDR) will be used meanwhile the rest hydraulic conductivity values for another cross section use appropriate values from the recent studies. The model is exported to the GIS so that the exact position of the longitude and latitude coordinates can be identified. The last step before conclusions and suggestions is model validation by ensuring that the recharge-discharge limit obtained is in accordance with the coordinates of the system when found in the actual location. Bore log and monitoring wells data area used also in the step of validation.

The data collection in this experimental research is based on secondary data and third data. The secondary data are supported by the previous study of the Jakarta groundwater basin. This includes many cross sections and some hydraulic parameter like the horizontal hydraulic conductivity (K 1) and vertical hydraulic conductivity (K 2). The third data that is produced by interpretation of bore log soil test that is also being enhanced in this research. Both the data type is utilized quantitatively. This research started in the beginning of February 2019 and the recharge discharge boundary area map is expected to be finish by the end of this year. This paper is published in order to support hypothesis of this research where stand at the middle of this research time frame. This 2 dimensional (2D) analysis methodology research is a new approach in finding a certain boundary area of recharge discharge in the Jakarta groundwater basin. The boundary area of recharge discharge in a horizontal map of Jakarta will be presented first time approximately at the end of this year.

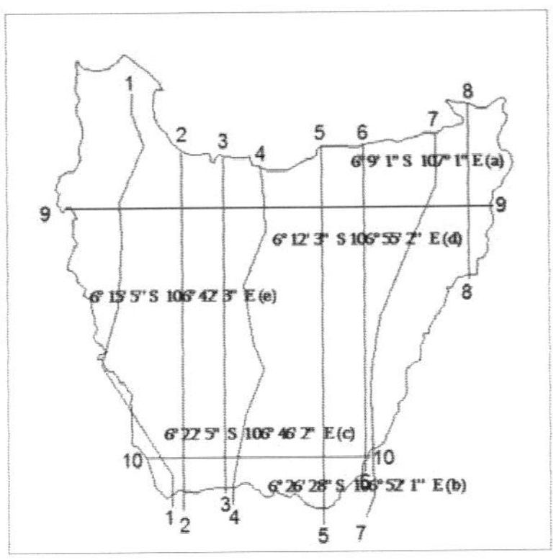

Figure 2. 10 Selected cross sections

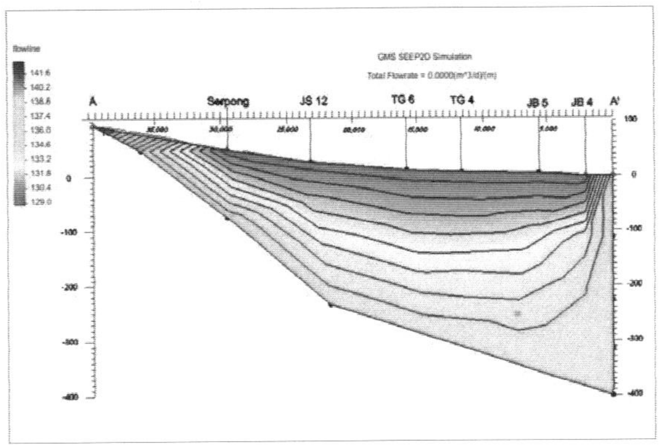

Figure 3. Cross section 1-1 Simulation in SEEP2D

4. Hypothesis and discussion

From the groundwater contour head map we selected 3 recharge areas and from the stratigraphy research indicates also 3 recharge areas in groundwater basin of Jakarta [3, 4]. The three coordinates with the highest head value resulted by the simulator is labeled with (a), (b), (c) in table 1. Each coordinate has hydraulic head more than 60 m. Otherwise, around 10 km to the south of Serpong Well, Bulok Kulon Well and Gunung Putri Well, the Stratigraphy study shows the area that coincides with impermeable formations that acts as a barrier for another groundwater water flow (no flow) from South, so these three areas also respond as a recharge area. In summary, from the simulator research and the stratigraphy study [3, 4] assume that there are 6 recharge area (table 1) hypothetically.

The 3 lowest hydraulic head those we can find in simulator area has less than 10 meters in head pressure value. These 3 cones depression are labeled with (d), (e) and the coast of North Jakarta. Since they have the lowest hydraulic head, the best hypothesis for these areas is a function of discharge area. Almost the same concept from the stratigraphy we will find in the map that all the water direct to Java Sea. In total there are 3 discharge areas (blue color) in figure 4.

Table 1. Hypothetical recharge – discharge boundary map location hypothesis

	Simulator Source	Stratigraphy Source
	Coordinate UTM / Area	Approximated area
Recharge area	(a) 6°9’16” S 107°1’6” E (Tambun, West Java)	≈ 10 km to south from Serpong well (Curug, Parung, West Java)
	(b) 6°26’28”S 106°52’12” E (Jatijajar, West Java)	≈ 10 km to south from Bulok Kulon well (Bojong, West Java)
	(c) 6°22’57” S 106°46’28” E (Sawangan, West Java)	≈ 10 km to south from Gunung Putri (Gunung putri, West Java)
Discharge area	(d) 6°12’37” S 106°55’29” E (Pulogadung, Jakarta)	Java Sea
	(e) 6°15’53” S 106°42’33” E (Perigi kulon, Banten)	
	The coast of North Jakarta	

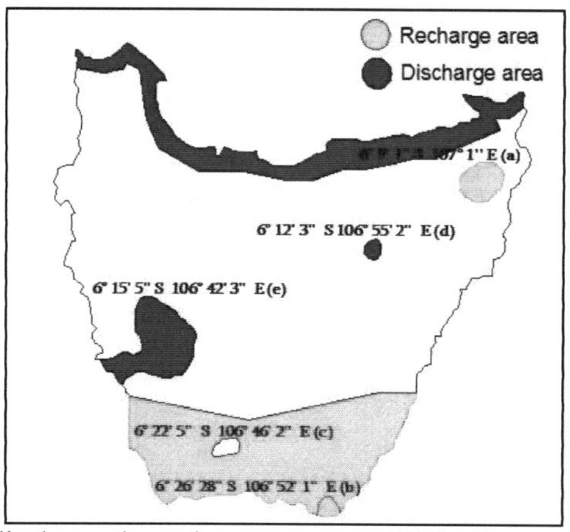

Figure 4. Recharge and discharge boundary area map of Jakarta groundwater basin hypothesis

There are two interesting phenomena also in figure 4. Discharge area (blue color) adjacent with recharge area (green color) in west south perimeter area. Another phenomenon that we can find is there is recharge area (green color) near to the discharge area (blue color) in east north of Jakarta groundwater basin territory. All phenomena eager us to hypothetically say that there are local groundwater systems in Jakarta groundwater basin. One of the example of the cross section is being simulated in this paper (figure 3). From that figure we can see after running SEEP2D (GMS) the result shows that regionally groundwater flow from south to north. This is evident compare to the real situation of Jakarta groundwater basin today and it is approved that the method that is being used in this research is eligible and appropriate to find the boundary area of recharge and discharge.

5. Conclusion

From theoretical consideration and paper review about Jakarta groundwater, the only thing that can be found is recharge discharge area map but not the recharge discharge boundary area map. This research is intended to determine the recharge discharge boundary area and to plot each area in a horizontal map. The finite element method in SEEP2D (GMS) program will be simulated in this research, including recharge and discharge boundary area map of Jakarta groundwater basin hypothesis (figure 4). Since the result of the investigation is not finished yet, this paper wants to demonstrate the methodology that is being conducted in the research and recharge discharge boundary area horizontal map hypothesis. In the future, this recharge and discharge boundary area horizontal map of Jakarta groundwater basin result will encourage many practical solutions in supporting Jakarta groundwater basin management.

6. References

[1] Freeze, R.A and Cherry, J.A. *Groundwater.* Prentice Hall, Englewood Cliffs, New Jersey (1979).
[2] D. E. Irawan, R. F. Lubis, H. Silaen, B. Brahmantyo, P. Sumintadireja dan D. J. Puradimaja. *Groundwater–surface water interactions of Ciliwung River streams, segment Bogor–Jakarta, Indonesia*, Springer-Verlag Berlin Heidelberg 2014
[3] Lambok M Hutasoit, Agus Mochamad Ramdhan dan Muhammad Fachri, *Stratigrafi dan Hidrostratigrafi Cekungan Airtanah Jakarta*, (2002)
[4] Riko Apriatresnayanto, *Pengembangan perangkat data untuk kebutuhan simulator pengelolaan airtanah pada Cekungan Airtanah Jakarta*, tesis (2018)
[5] Fetter, C.W, *Applied Hydrology* Prentice – Hall, (2001)

Acknowledgment

The Author would like to thank The University of Indonesia in supporting this research through Indexed International Publication Grants for Student Final Project (Hibah PITTA 2019). The grant is used in publishing this paper, accommodation and some other expanses in the process of finishing this paper.

IOP Conf. Series: Materials Science and Engineering **690** (2019) 012009 doi:10.1088/1757-899X/690/1/012009

Waste Management System of Badung River in Bali Based on Community Behaviour

IGAI Mas Pertiwi[1], **W S Kristinayanti**[1], **W Sudiasa**[1] **and K W Andayani**[1]

[1]Civil Engineering Departement, Bali State Polytechnic, South Kuta, Badung Regency, Bali 80361

Email: maspertiwi72@yahoo.co.id

Abstract. The identification of garbage at several points along the Badung river shows a low level of public awareness of the environment. The origin of the contents of the garbage in the Badung river comes from various human activities such as industry, household and nature. Although there is a trash rack, but it is not optimal in reducing plastic waste carried on the river. There are several reasons that cause people to throw garbage into the river. Among other things, disposing of garbage into the river is considered more practical and freer, the lack of garbage disposal facilities around the river and has become a culture. Therefore, research was conducted on community behaviour in disposing of garbage along the Badung river. The results showed trigger factors variable with a very influential factor, namely the existence of a disease of 79%. And as many as 5% of the people feel that the availability of Temporary Disposal Site facilities has no effect. The waste management system will be managed directly by the Waste Self-Management Group in each village, as an effort to build an integrated system. It also educates the public how to sort, manage and utilize waste. With the proper governance, of course, it can reduce the volume of waste to landfill.

1. Introduction

1.1. Background

The arrangement of Badung river has begun, but public awareness of the importance of waste management has not been fully realized with the visible piles of garbage in several places along the Badung river flow. Garbage is a sight that is still found along the river that crosses Denpasar City. Residents outside the watershed also contribute to the dirty Badung river. Need education in the community so as not to throw garbage into the river. A few years back the Badung river had experienced siltation, so when the rainy season arrived, the river often overflowed [1]. Previous study provides evidence that household and community groups' waste disposal practice is careless with the environment. Such waste disposal practice with disregard for the possible environmental consequences is possibly influenced by specific socioeconomic status (sex, education attainment, and household income) and geographic risk factors (residential area and residents' distance to municipal permitted dumpsite). It demonstrated that the respondents not only have poor knowledge of the adverse health effect with regard to improper waste handling but also have unsafe behaviour towards safety practices [2]. To examine the influences of internal and external factors on people's waste disposal behaviours in Phnom Penh, the results showed that personal and social norms, such as perception of social pressure from friends and family and from the government, had significant influences on intention, whereas the influences of external factors was much smaller [3]. Lack of public awareness and coordination between agencies involved and other bad natural phenomena that might occur certainly can have a big influence on the environment, natural scenery, conservation of coastal waters, water availability can even cause more severe ecosystem damage along Badung river.The integration of the handling program needs to be prepared and matured and applied well in order to create a clean, safe, sustainable and beautiful ecosystem. Identification of community behaviour in disposing of garbage is very necessary in order to develop a waste management model and this arrangement becomes more important to do besides because Denpasar is the provincial capital also because the Badung river is planned as one of the city's tourist attractions.

1.2. Objectives

The objectives of this study are:

1. Identify the dominant factors that cause community behaviour in disposing of garbage in the area along the Badung river catchment area.
2. Designing the right waste management system to overcome the problem of waste in the Badung catchment area.

2. Research methods

2.1. Research sites

The object of this research was carried out along the Badung river catchment area, namely by analysing 8 survey location points, where 8 points of this location were the river arm which empties into the main river of Badung river namely, Badung river, Langan river, Medih river, Jurang river, Guming river, Ketapian river, Kelandis river and Sudetan Teba river.

2.2. Identification of community behaviour in dispose of trash

The method of observing and retrieving data on the behaviour of people disposing of garbage along the Badung river watershed is by asking a number of questions to respondents who will be answered according to the questions in the questionnaire. The scale used in this study is the Guttman scale [4]. The assessment category is very influential, not influential and not very influential.

3. Result and discussion

3.1. The socio-economic condition of Denpasar City

Denpasar, administratively, has 4 sub-districts and 43 villages. South Denpasar District consists of 10 villages. East Denpasar 11 Villages / Villages. West Denpasar 11 Villages / Villages. Denpasar Utara 11 Village. The total area of Denpasar City is 12,778 hectares or 2.18% of the total area of Bali Province.[5]

3.2. Demographics

The population increase of Denpasar City as the provincial capital is not only influenced by the natural growth of the population but also the high flow of migration. This situation is reflected in the high population density of Denpasar. Based on the results of population projections, the total population of Denpasar City in 2015 was 659,623 people. With such a large population, the population density in Denpasar City in 2015 reached 6,891 people per km^2. If seen by age group, the population of Denpasar in 2015 is dominated by productive age population (ages 15-64 years).[5]

From the data issued by almost 42.19% of Denpasar residents live around the Badung river, so that if not done a good management it will have a significant problem.

3.3. Education

In 2018, Denpasar City has 226 primary schools. at the junior secondary level there are 63 schools, and at the high school/vocational level, there are 65 schools.

From the data it can be seen that every year the level of education of the citizens of the city is increasing, where the percentage of citizens who have high school education or who are equal is

85.94%. With a good level of education, it should be easier for the government to sensitize Denpasar residents about the importance of the river, so that what needs to be improved is the ease of the community in carrying out garbage disposal.

3.4. Areas identified as trash along Badung River

This research focuses on the waste problem, where in the field searches that have been carried out along Badung river, identified at several points along the Badung river there are garbage, seen on figure 1 bellow:

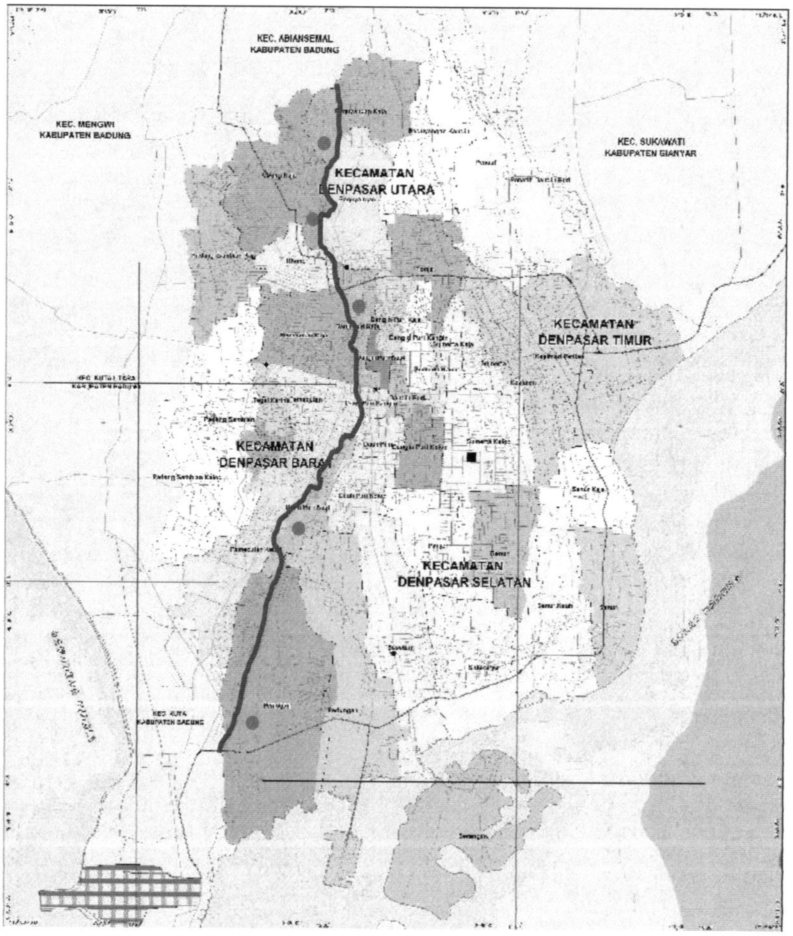

Figure 1. Areas identified as trash along Badung River

3.5. Community behaviour survey

The river as a cultural centre of society in the past has now lost its function, the river as a centre of community activity can no longer be found at this time, a river by some people is used as a backyard that lacks attention. At present the river is considered a less useful place so that people are accustomed to and assume that it is permissible to dispose of garbage in their backyard (river). The community makes the river a landfill. There are several reasons that cause people to throw garbage into the river. Among other things, throwing garbage into the river is considered more practical and freer, the lack of garbage disposal facilities around the river and has become a culture.[6]

3.6. Population and samples

In this study the population is households in residential areas within the area along the object of the research, namely the Badung River Catchment area, based on a preliminary survey that identified waste. The population of the survey area is 51478 households. Samples to be taken from the population must be truly representative or representative. In determining the sample, researchers used a withdrawal technique with the Simple Random Sampling method. This method can be done if members of the population are considered homogeneous.

3.7. Data collection

Data collection was done by distributing questionnaires by 5 surveyors in each sub-district by dividing the number of samples according to the results of the above calculations, namely the districts of North Denpasar and West Denpasar as many as 30 samples and South Denpasar sub-districts as many as 40 samples. Questions in the questionnaire consisted of 4 variables, namely background, belief and mental readiness, facilities and trigger factors. Each variable has 2-3 factors that are expected to influence people's behaviour in disposing of garbage. Questionnaires were made with scale 3 questions to find out whether these factors were very influential, not influences and were not very influential on people's behaviour in disposing of garbage. The variables and their factors are as follows:[7]

Table 1. Variable of Community Behaviour and Its Factors

X1 = background	X2 = mental trust and readiness	X3 = facilities	X4 = trigger factor
X11 = education, occupation, salary	X21 = the advantage of keeping the river clean	X31 = temporary disposal site	X41 = disease
	X22 = loss of using dirty water	X32 = kitchen site	X42 = fine
X12 = culture	X23 = garbage and self-cleaning activities shouldn't be thrown into the river	X33= self-cleaning activities site	X43 = counseling

3.8. Data analysis

Data analysis using tabulation method using Microsoft Excel software by giving a score of 1 to the choice of respondents. The results of the analysis of community behaviour in disposing of waste are then labelled based on the choice of answers on a scale of 3, which is very influential, not influential and not very influential. The results of the analysis of community behaviour can be seen in the following graphs:

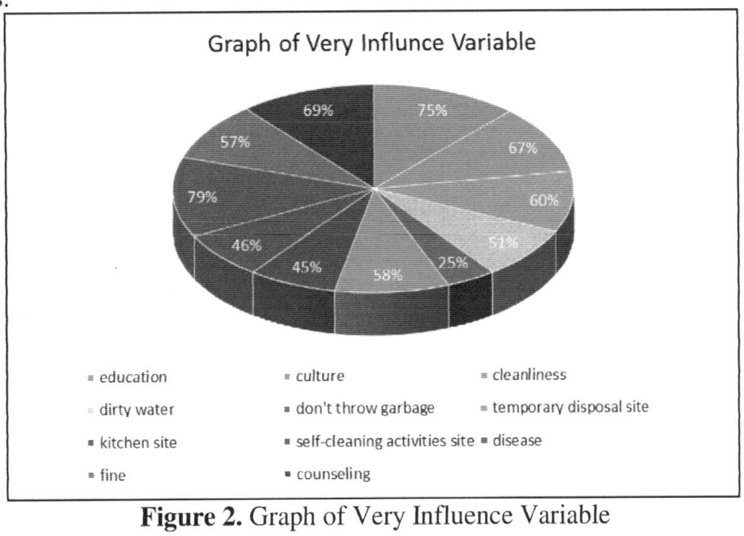

Figure 2. Graph of Very Influence Variable

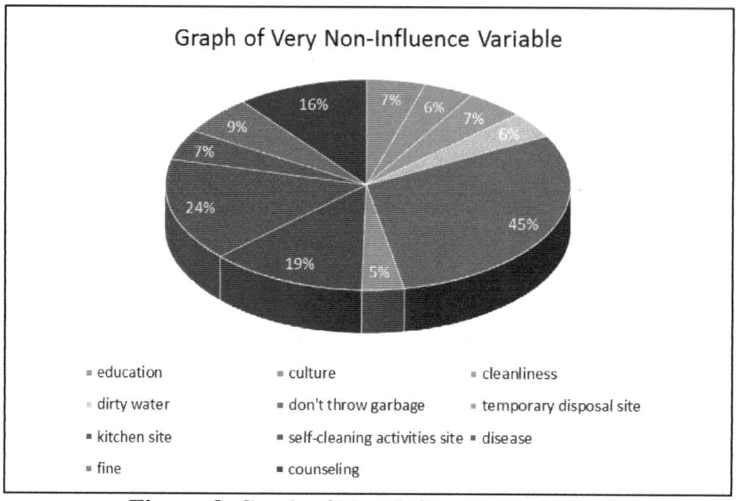

Figure 3. Graph of Non-Influence Variable

Based on figure 2, it can be seen that the variables that greatly influence the behaviour of people in disposing of garbage are trigger factors variables with very influential factors, namely the existence of a disease of 79% variable followed by X1, namely background with very influential factors are education, employment and income of 75% and counselling of 69%. This shows that education and employment greatly affect the community to dispose of garbage in its place but have little influence on the belief not to dispose of garbage and self-cleaning activities into the river.

Based on figure 3, it can be seen that the variables of trust and mental readiness which have a percentage do not affect the trust of losses using dirty river water by 43% followed by the variables of the means of Temporary Disposal Site and kitchen location of 37% and 36%. This shows that the existence of a disease does not affect the community in disposing of garbage but with a very small percentage, while 43% of the community feels that the loss of using dirty river water does not affect people's behaviour in disposing of garbage.

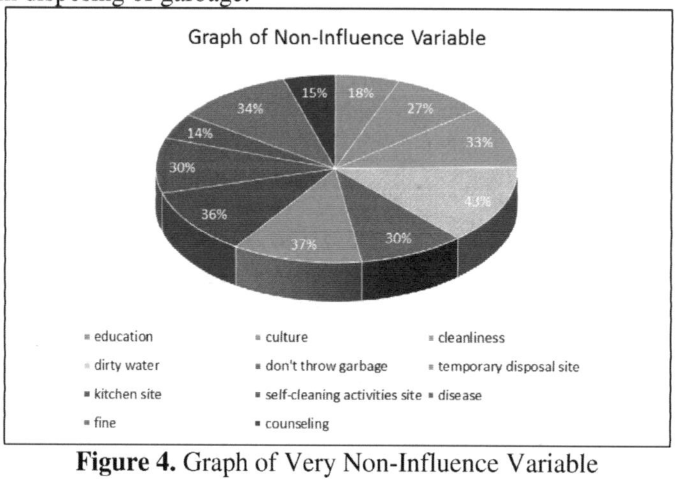

Figure 4. Graph of Very Non-Influence Variable

And based on figure 4, it can be seen that the variables of trust and mental readiness, namely the public's belief that garbage and Self-Clean Activities may be disposed of into the river, do not affect the community to dispose of garbage into the river by 45%. This shows that still some people still adhere to the belief that it is very not influential to dispose of garbage and self-cleaning activities into the river, while the availability of Temporary Disposal Site facilities does not really affect the community to dispose of garbage in its place.

IOP Conf. Series: Materials Science and Engineering **690** (2019) 012009 doi:10.1088/1757-899X/690/1/012009

3.9. Waste management system model plan

The management of municipal solid waste management is getting more severe due to various reasons such as poor land use and infrastructure, weak technical and financial capacity, lack of enforcement of regulations, poor coordination between authorities, deficient policies and absence of political priorities. Municipal solid waste management starts with understanding public concerns, preferences, knowledge and behaviour. The most cost-effective way of reducing municipal solid waste include public education and citizen encouragement to share in the design of recycling processes. Moreover, public participation strongly affects the success of recycling processes [8].

The maximum sorting and recycling activities are carried out from the garbage collection to the final waste disposal. Operational techniques for urban waste management consisting of storage activities up to the final disposal of waste must be integrated by sorting from the source.

First of all, most people consider recycling to be a low priority task and tasks with lower priority have a tendency to not be performed if the task is perceived as inconvenient. Therefore, it is important to design recycling facilities as close to the recycler as possible and facilities should be equipped to work in a given situation, for example, with lower access points. Secondly, it is important to have knowledge about how and where to recyclers or non-recyclers tend to have less or obsolete information that prevents them from recycling. Even if motivation and understanding are in place, without knowledge of how to recycle, it still does not work. Finally, a contributing step is to establish a deeper understanding of the environment and the impact of recycling to gain or maintain a general level of motivation, which might further increase recycling [9]. Based on the results of the research above, the analysis of community behaviour in disposing of garbage results in education and employment greatly influencing the community to dispose of garbage in its place but has little influence on the belief not to dispose of garbage and self-cleaning activities into the river and some people still believe that it does not affect the disposal of garbage and self-cleaning activities into the river, while the availability of Temporary Disposal Site facilities does not significantly affect the community to dispose of garbage in its place. And based on the type of garbage observed visually, the dominant organic waste is obtained from vegetables and wet fruits from kitchen waste. Whereas the dominant an organic waste is food packaging and plastic bags. From these results and based on Denpasar City Governor No. 11 of 2016 [10], the procedures for managing and disposing of waste in Denpasar City are supposed to be carried out by waste management groups coordinated by the village as shown in figure 5. So that the handling of waste starting from upstream can provide education to the community about the sorting, management and utilization of waste. With proper governance, it can certainly reduce the volume of waste to landfill, and inorganic waste can be sorted and saved in a Waste Bank, while organic waste can be used for compost.

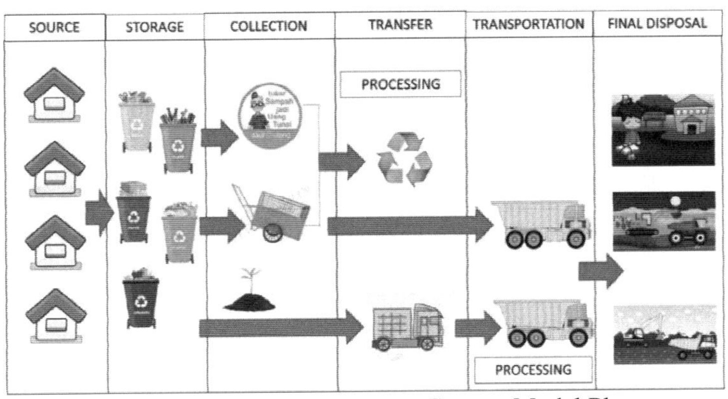

Figure 5. Waste Management System Model Plan

4. Conclusion

Based on the results of the research above, the analysis of community behaviour in disposing of garbage results in education and employment greatly influencing the community to dispose of garbage in its place but has little influence on the belief not to dispose of garbage and self-cleaning activities into the river and some people still believe that it does not affect the disposal of garbage and self-cleaning activities into the river, while the availability of Temporary Disposal Site facilities does not significantly affect the community to dispose of garbage in its place. From these results and based on Denpasar City Governor No. 11 of 2016, the procedures for managing and disposing of waste in Denpasar City are supposed to be carried out by waste management groups coordinated by the village. So that the handling of waste starting from upstream can provide education to the community about the sorting, management and utilization of waste. With proper governance, it can certainly reduce the volume of waste to landfill, and inorganic waste can be sorted and saved in a Waste Bank, while organic waste can be used for compost.

5. References

[1] https://www.cendananews.com/2018/06/700-kg-sampah-diangkut-setiap-hari-dari-sungai-tukad-badung-denpasar.html. Bali Post

[2] Keita Mamady, 2016. *Factors Influencing Attitude, Safety Behavior, and Knowledge regarding Household Waste Management in Guinea: A Cross-Sectional Study*. Journal of Environmental Public Health. 2016; 2016: 9305768. Published online 2016 Mar 22. doi: 10.1155/2016/9305768

[3] Pagnarith Srun, Kiyo Kurisu, 2019. *Internal and External Influential Factors on Waste Disposal Behaviour in Public Open Spaces in Phnom Penh, Cambodia*. www.mdpi.com/journal/sustainability,11,1518; doi: 10.3390/su11061518.

[4] Sugiyono, 2010. *Quantitative, Qualitative and R&D Research Methods*, CV.Alfabeta, Bandung.

[5] BPS-Statistics of Denpasar Municipality, 2018, *Denpasar Municipality in Figures*, BPS-Statistics of Denpasar City.

[6] Setiawan, 2009, *Martapura River is The Source of Life*. http://sebuah blog.blogspot.com/2009/02/sungai martapura sumber kehidupan. Html.

[7] Liana Penny, 2012, *A Study of People's Behaviour to Dispose of garbage on The Banks of The Martapura River on The Aquatic Environment*. EnviroScienteae 8 (2012) 117-126. ISSN 1978-8096.

[8] Gamze Turan, 2016. *The Importance of Environmental Education on Attitudes and Behaviours for Household Waste Management in Black Sea Region, Turkey*. The Eurasia Proceedings of Education & Social Sciences (EPESS) ISSN: 2587-1730. Volume 5, Pages 227-231.

[9] K.Johansson, 2016. *Understanding Recycling Behaviour: A Study of Motivational Factors Behind Waste Recycling*. Proceeding of International Conference on Waste Management and Environment (WM 2016). www.witconferences.com

[10] Perwali Kota Denpasar No. 11 Tahun 2016, *Procedures for Management and Disposal of Garbage in The City of Denpasar*. Denpasar.

IOP Conf. Series: Materials Science and Engineering **690** (2019) 012010 doi:10.1088/1757-899X/690/1/012010

Thermodynamic Prototype Research of Vernacular Architecture Based on Climate Adaption

Simin Tao[1] and Huanyan Chen[2]

[1] College of Architecture and Urban Planning, Tongji University, 1239 Siping Road, Shanghai, China
[2] Arcplus Group PLC, East China Architectural Design & Research Institute, 321 Sichuan Road, Shanghai, China

Email: tetcato@163.com

Abstract. Climate adaptation represents the thermodynamic state of building systems by feedback of form, material and spatial organization. This article proposes a new method to research the climate adaptability of vernacular architecture. Taking Shentan Village settlement as a case study, we search for the architectural prototype from the perspective of thermodynamics by simulations of LADYBUG+HONEYBEE tools, extract the architectural elements related to the specific climate characteristics, and systematically analyse the internal logic of energy flow in vernacular buildings by energy systems language.

1. Vernacular architecture and thermodynamic prototype

The contemporary evolution of energy and thermodynamics provides a special perspective for architectural autonomy and climate adaptation problem[1]. Kiel Moe put forward the concept of "energy formation" in his book *Convergence: an Architectural Agenda for Energy* in 2013, and explicated materials, energy systems and amortization as three factors that should be converged through design to maximize the architectural and ecological power of buildings[2]; in 2015 William W. Braham explored the architectural implications of system ecology in *Architecture and Systems Ecology*, extended the principles of thermodynamics with the resilient self-organization of ecosystems by means of energy systems language; in their book *Essays on Thermodynamics, Architecture and Beauty*, Inaki Abalos and Renata Sentkiewicz presented the arguments and designs around the new category of "thermodynamic beauty"[3]. For vernacular architecture, due to the relatively primitive local materials and construction technique, buildings must be simple but efficient. By studying the vernacular architecture in different climatic conditions, extracting their form elements and energy flow modes, we can understand the historical experience by people adapting to climate through buildings, that is, thermodynamic prototype.

2. Research method of vernacular settlements from the perspective of thermodynamics: a case study of Shentan Village in Yiwu

Shentan Village in Yiwu is located in Zhejiang Province, China. It belongs to the “hot summer and cold winter” region of Climate Regionalization of Architecture in China, characterized by distinct seasons and heavy rainfall. The local settlement in the village is situated on a slope, consists of typical vernacular dwellings with courtyards in Zhejiang.

2.1. Visualizing Climate Data with Parametric Performance Software

The LADYBUG + HONEYBEE Tools based on Grasshopper have the ability to customize the visualization of simulation results and the target time period to be analyzed. LADYBUG can be used to generate diagrams of temperature, humidity, wind roses, solar radiation, etc., and generate psychrometric chart to show the effectiveness of passive strategies. HONEYBEE can simulate environmental performance such as energy consumption and thermal comfort calculation. By inputting the climate date in EPW file of Shentan village into LADYBUG (Figure 1, Figure 2), the related climate data will be visualized and analyzed.

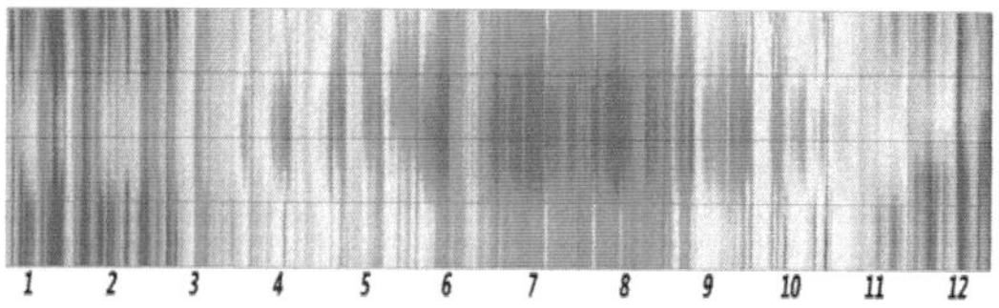

Figure 1. Dry Bulb Temperature of Shentan-Hourly (Legend: 0℃-35℃)

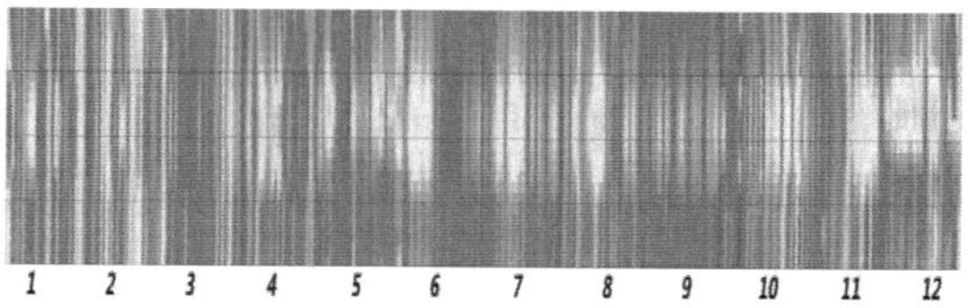

Figure 2. Relative Humidity of Shentan-Hourly (Legend: 20%-90%)

2.2. Analyzing energy demand based on climate characteristics

Through data visualization we can intuitively understand the climate characteristics, whether annual or seasonal, occasional or general. According to the analysis results of L+H tools, the important climatic data of Shentan Village are as follows (Table 1). The table shows that the temperature of Shentan Village varies greatly throughout the year. The annual total precipitation and humidity are high and makes people necessary to pay attention to rain-proof and moisture-proof in structure and material. Radiation resource is rich and should be fully utilized. The most significant problem in energy demand is how to deal with the contradiction between heat insulation in summer and heat preservation in winter.

Table 1. Major Climate Data of Shentan Village

Avg. monthly min. temp. (℃)	Avg. monthly max. temp. (℃)	Min. temp. (℃)	Max. temp. (℃)	Average relative humidity (%)	Solar elevation angle on June 21 (°)	Solar elevation angle on Dec. 21 (°)	Annual precipi-tation (mm)	Avg. daily total solar radiation (Wh/㎡)
4.5	27.2	-14.5	36.2	77.6	84.2	37.4	1300	4900

2.3. Summarizing the energy strategies of vernacular architecture

According to the demand, the energy strategies of vernacular architecture can be analyzed from the following aspects: (1) reasonable location and layout; (2) adapt to the topography and actively utilize existing local construction materials; (3) heating with natural heat source and cooling with natural heat sink by rational organization; (4) programming the functions to achieve energy coordination of various indoor spaces; (5) with certain low-tech ways to create better thermal environment conditions. Based on field research and measurement of Shentan settlement including assessment of indoor physical environment and infrared thermal imaging, the energy strategies of vernacular dwellings to cope with climate were summarized as follows:

- Street space with intensive road network: shading and ventilation;

The widest main road in Shentan Village is only 4m, and the aspect ratio of street space is more than 2:1. According to the sunlight hour simulation by L+H tools (Figure 3), the sunshine inside Shentan is far less than outside open space due to mutual occlusion. Open spaces such as ponds and squares in village form good conditions for wind pressure ventilation.

- Patio with appropriate scale: lighting, ventilation and drainage;

Patio is a prominent spatial feature of vernacular architecture in Zhejiang Province, an important way of lighting the interior rooms. Due to the strong solar radiation and the stack-ventilation function, the size of the patio is relatively narrow. Also there are kennels inside the patio to guide rainwater from indoor to outdoor drainage ditches.

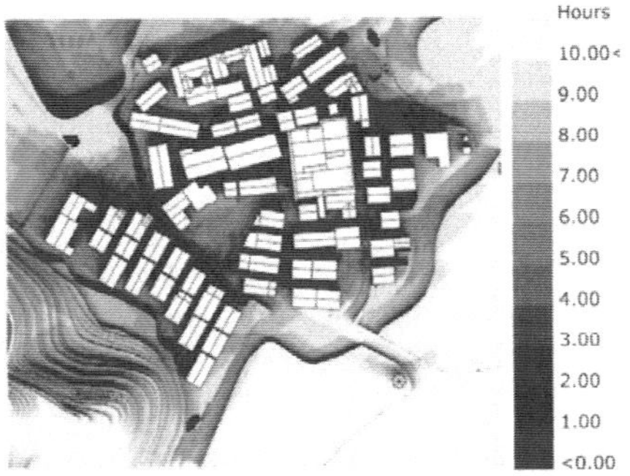

Figure 3. Sunlight hour analysis of vernacular settlement in Shentan

Figure 4. Thermograph of the patio in Feng Xuefeng's former residence

Figure 5. Enclosed envelope

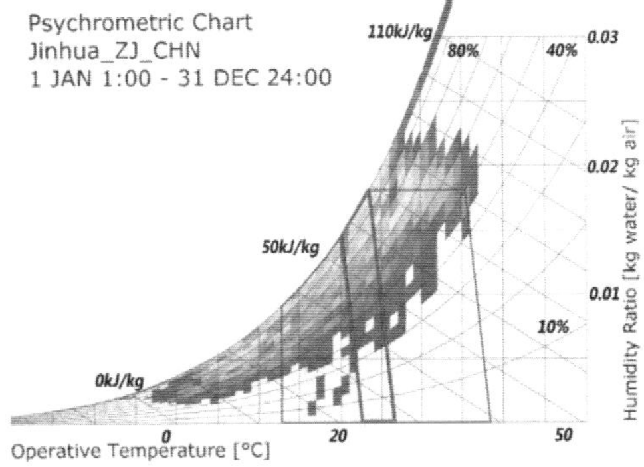

Figure 6. Comfort zone before and after passive strategy comparison(0~120h)

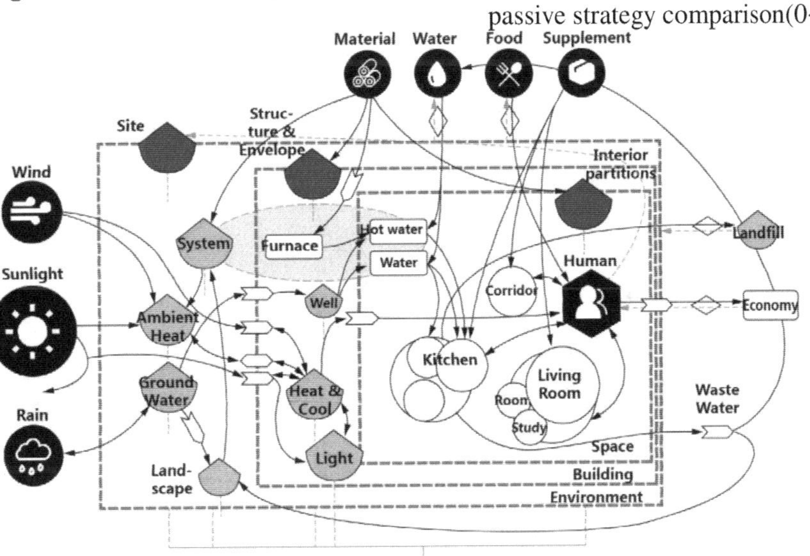

Figure 7. Energy systems language diagram of vernacular dwellings in Shentan Village.

- Insulated envelope: thermal storage;

Wood and stone are abundant in the middle of Zhejiang Province, and the construction materials of vernacular architecture are basically draw on these local resources. The exterior wall is mainly made of materials with good thermal stability and inertia, such as brick wall and earth wall, which are adapted to the variable local climate.

- Sloping tiled roof: thermal storage and rain-proof;

It can be seen that the tiled roof of Feng Xuefeng's former residence has a strong heat storage capacity from thermographs (Figure 4). In winter night, the temperature of the roof surface can be higher than elevation surface by more than 5℃. In addition, the double eaves slope roof guides rainwater to the kennels inside the patio to prevent rain from scouring the wall.

- Corridor space: buffer zone;

Measurement data shows that the air temperature in the corridor on second floor is always between patio and rooms (Figure 5). The corridor is another buffer zone between indoor and outdoor space, which stabilizes the amplitude of indoor temperature.

- Compact centripetal layout: reducing heat loss;

Feng Xuefeng's former residence is one of the most representative local dwellings in Shentan Village. It is a quadrangle with rooms arranged compactly around the central square patio. Relative to the exterior enclosed wall, the interior rooms facing the patio have light wooden plank partitions toward the patio. The compact centripetal layout prevents the influence of external environment fluctuation on the internal space.

Feedback their passive strategies to L+H software for thermal comfort simulation can verify which passive strategies are more effective and become the key link in architectural thermodynamic system. The indoor thermal comfort zone of Shentan vernacular dwellings can be increased from 4.6% to 37.6% (about 3298h) by the strategies above (Figure 6).

3. Architectural thermodynamics prototype method based on climate characteristics

The research of vernacular architecture types from the perspective of thermodynamics involves a comprehensive influence of natural environment factors. It summarizes how the climate strategy is expressed through types in specific sites and how the prototype be reconstructed.

3.1. According to the basic types of energy transfer modes

Energy flow from renewable resources such as wind, sunlight, heat and rainwater always circulates in the open system between buildings and environment, and buildings utilize or block energy flow from the environment through spatial organization, material construction and technical equipment[4]. The energy transfer structure in vernacular architecture can be simplified into three modes: energy capture, energy programming and energy regulation, it is the basis for prototype extraction. The main ways of energy capture mode include radiation, conduction and convection, and the object is the natural energy outside the buildings; energy programming mode means the heat sources and heat sinks in buildings are organized carefully to form a good cooperation based on the energy balance; energy regulation mode is an auxiliary structure to control natural energy or internal heat gain by means of low-tech and low-cost methods[5]. Take vernacular dwellings in Shentan Village for graphical analysis in energy systems language, the energy and material flows in the system will be presented in three scales: space, building and environment (Figure 7).

3.2. According to energy demand types based on climate characteristics

External climate factors related to vernacular architecture include air temperature (T), diurnal temperature difference (D), humidity (H), precipitation (P), wind (W) and solar radiation (R), different elements correspond to different energy requirements: heat storage (+) or heat loss (-) according to (T); need (+) or no need (-) to control indoor temperature fluctuation according to (D); humidify (+) or dry (-) according to (H); storage water (+) or drain off water (-) according to (P); ventilation (+) or windproof (-) according to (W); increase or reduce heat gain according to (R). The factors affecting energy demand also include indoor heat sources (S), whether to increase (+) or decrease (-) heat production, and whether the heat produced needs to be accumulated (+) or dissipated (-). The final results can be presented as a list. Different energy demand types directly lead to the climate strategies of a building, which is reflected in the architectural form.

3.3. According to energy strategy types of different architectural element scales

Energy prototype elements can be divided into five levels: group layout, plane level, vertical level, interface porosity and auxiliary technology. These elements establish an effective link between climate conditions and human comfort. Thermodynamic prototypes of vernacular architecture in different cases can be analyzed in a graphical way in combination with climate characteristics. Table 2 shows the thermodynamic prototype comparison of four cases of vernacular settlement from different climatic zones in China.

Table 2. Thermodynamic prototype of four different vernacular settlement

Settlement	Climate	Group layout	Plane level	vertical level	Interface porosity	Auxiliary technology
Shentan Village, Yiwu, Zhejiang	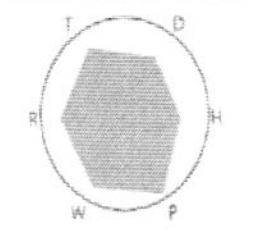		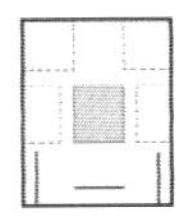	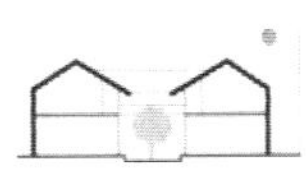	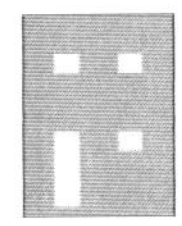	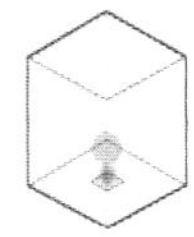
Hushan Town, Quanzhou, Fujian	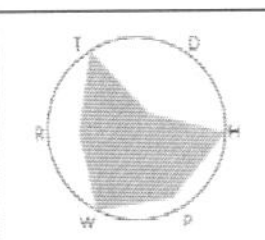		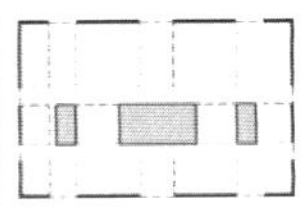	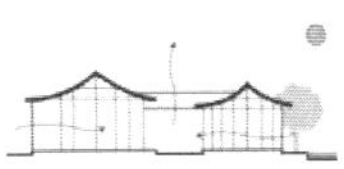		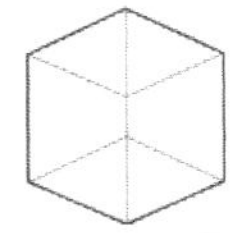
Baishe Village, Sanyuan, Shanxi	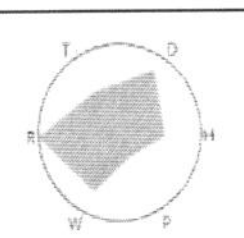	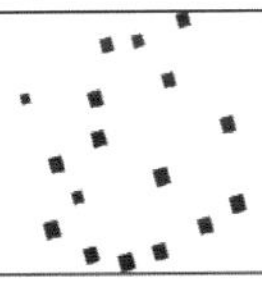	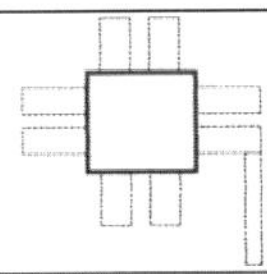	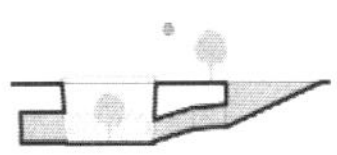		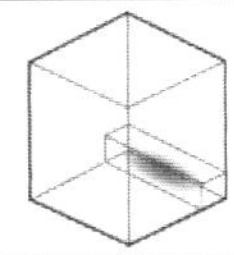
Yaozhan Village, Fushun, Liaoning	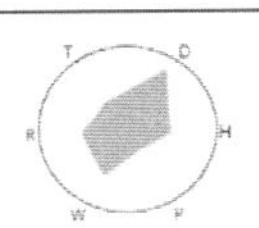		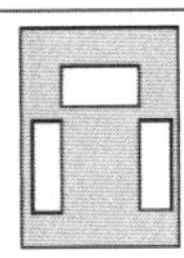	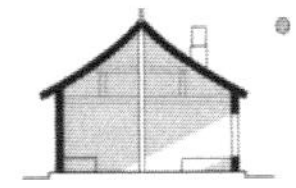		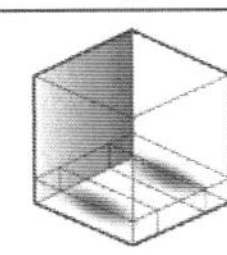

4. Conclusion

To study the process of thermodynamic prototype of vernacular architecture, it is necessary not only to analyze the relationship of energy I/O between environment and buildings, but also to consider the organization of energy flow in building systems[6]. The thermodynamic prototypes can be regarded as a reference to study the logic of architectural spontaneous generation in specific climate, and applied to contemporary architectural design.

5. References

[1] Li L and Tao S 2016 *South Architecture* **173** 29
[2] Moe K 2013 *Convergence: An Architectural Agenda for Energy* (London &New York: Routledge) pp 124-125
[3] Abalos I 2014 *AV Monographs* **169** 4
[4] Braham W W 2015 *Architecture and Systems Ecology: Thermodynamic Principles of Environment Building Design* (London &New York: Routledge) pp 27-30
[5] Abalos I and Snetkiewicz R 2015 *Essays on Thermodynamics, Architecture and Beauty* (New York: Actar) p 89
[6] German J G 2017 *Thermodynamic Interactions: An Exploration into Physiological Material and Territorial Atmospheres* (New York: Actar) p 151

Acknowledgment

The paper is supported by Key Laboratory of Ecology and Energy-saving Study of Dense Habitat (Tongji University), Ministry of Education of China.

IOP Conf. Series: Materials Science and Engineering **690** (2019) 012011 doi:10.1088/1757-899X/690/1/012011

A Study of Architectural Design of Thai Temples Influenced by Thai and Mon Ethnics in Sam-Khok District

M.L.Varodom Suksawasdi

Faculty of Architecture and Planning, Thammasat University, Pathum Thani, Thailand, 12121

Email: varodom@ap.tu.ac.th

Abstract. This study aims to investigate and compare architectural design including layout planning and stupa design of Thai and Mon Buddhist temples in Sam-Khok district. The data collection is gained from literature reviews, field observation, and interview. It is found that the design of these two temples is slightly transformed from those of the royal temples in terms of the layout axis and size of stupa. The change of the main access from water-based to land-based approach effects the layout of Buddhavas and Sangkhawas zones. These findings can provide an understanding of influential factors on architectural design of Thai and Mon Temples. It is benefiting the cultural tourism development in Sam-Khok district.

1. Introduction

"Sam-Khok district" known as 'Mueang Sam-Khok' is an ancient city of Thailand [1].The north part is connected to the old town Ayutthaya province. The city was established in early Ayutthaya period. King Narai provided the land for the settlement of Mon migrated from Burma. The Mon is famous for their expertise in potteries [2]. Besides the potteries, Sam Khok has high opportunities for cultural tourism development since this area consists of several temples and local community. It has rich local history, arts, and cultures [3]. However, information of architectural design of temples in Sam-Khok has not been investigated. This study aims to investigate an impact of Thai and Mon Buddhist Ethnics on temple layout and architectures in Sam-Khok district. The data is assessed by using contextual analysis method as a preliminary field investigation to analyze the architectural design process. The condition of the site such as building's layout, zoning, architectural elements and style is carefully studied to project the potential of the site in the future. [4].

2. Layout and architectural design of Thai temple

The compound of Thai temple layout is typically divided into three main sections: Buddhavas, Sangkhawas, and Sangha. The Buddhavas is the most important area that the monks use for religious ceremonies. Sangkhawas area is the residential area for the monks. It consists of a monk residential building and other personal facilities called 'Kuti.'. The Sangha is the area for public use such as school and crematorium (Phra-meru). This study focuses only a relationship between Buddhavas and Sangkhawas areas. Sangha area is excluded from the study. Previous study investigated the layout and location of Buddhavas and Sangkha was from several temples constructed in Ayuthaya and Ratanakosin periods, which can be separated into 4 main layouts as shown in Figure 1. Typically, the Buddhavas area is located on the area with easiest access from the neighboring communities while Sangkhawas area is located in private zone.

Therefore, Buddhavas is the most sacred place where the monks conduct religious ceremonies. This area consists of three architectural elements, namely stupa (Chedi) in (a), chapel (Bot) in (b), and sanctuary (Viharn) in (c) as presented in Figure 2. Table 1 presents typical layout of Buddhavas with those three architectural elements located along different axes [5].

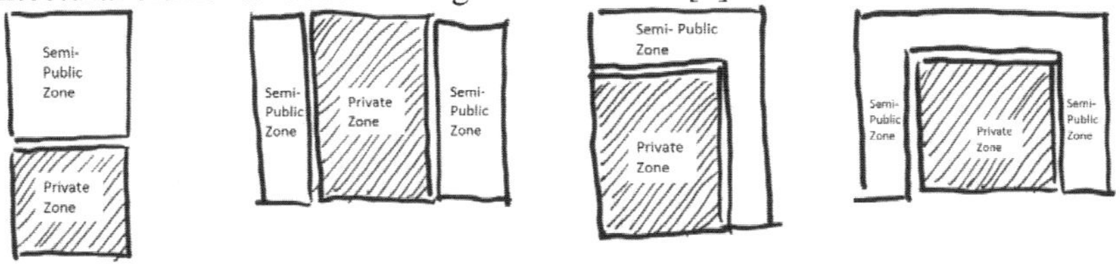

(a) Single side (b) Separation (c) Elbow (d) Courtyard

Figure 1. Design of Buddhasawas and Sangkhawas locations in the compound of Thai temple layout

(a) (b) (c)

Figure 2. Architectural design of stupa (a), chapel (b), and sanctuary (c).

Table 1. A layout planning of main stupa (circle), chapel (shaded square), and sanctuary (unshaded square) found in different periods [5].

Era (year)	**Types of temple's layout**				
	Single axis	Double axis	Tripple axis	Axial layout scene	Cross axis
	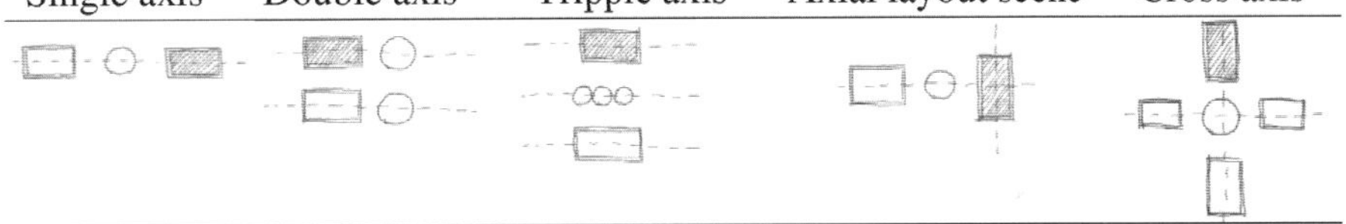				
Tawaravadee (B.E.12 -16)	✓	×	×	×	×
Sukhothai (B.E.19 - 20)	✓	✓	✓	×	×
Ayutthaya (B.E.20 - 23)	✓	✓	✓	✓	✓
Rattanakosin (B.E.23 - present)	✓	✓	✓	✓	✓

3. Research methodology

Firstly the study selects two temples, that their architectural designs are influenced by two different ethnic groups living in Sam-Khok district the history of the temples and local communities is reviewed from historical documents and former studies. The contextual analysis emphasizes on the design of the layout arrangement of the architectural elements in each zones. The arrangement of such elements is collected by field investigation and the interview with the abbot. Then the temple's zoning, architectural layout, and stupa style of those two temples are compared and discussed (Figure 3).

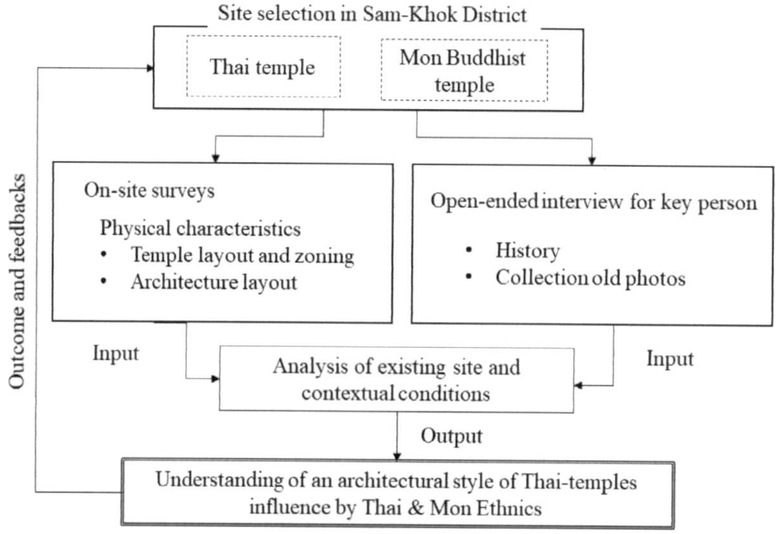

Figure 3. A framework of this study

As for the case study selection, there are 30 temples situated in Sam Khok district. This study chooses Wat Thour Thong (a) and Wat Sam Lae (b) as the case studies because the architectural design of Wat Sam Lae represents Mon Buddhist temple while Wat Thour Thong represents Thai Buddhist temple. These two temples were built in early Rattanakosin period (during 1824-1837) and both locate on the west bank of Chao Phraya River (c) as shown in Figure 4.

(a) (b) (c)

Figure 4. Selected case studies: Wat Thour Thong (a) and Wat Sam Lae (b). (c) presents the location of the two temples.

4. Results

According to historical evidences, both temples were built by the local communities, but difference in ethnic groups. The funding and construction contribution to built Wat Thour Thong was made by Thai Buddhist while the Mon funded and constructed Wat Sam Lae. Wat Thour Thong used to be a transportation hub and marketplace where people traded and exchanged products. Since the location was easily accessed for the monks, people later built this temple for monk's residence and serving religion ceremonies. Wat Sam Lae was the place where the Mon emigrated from Burma, settled and formed the community in Sam-Khok after the nine-armies wars in King RAMA I era. The activities in Wat Thour Thong were conducted under Maha Nikai lineage principles while the activities in Wat Sam Lae were conducted under Dhammayuttika Nikaya principles [4]. The layout and architectural designs of these two temples are compared and discussed according to the following topics:

4.1. Layout planning

Figure 5 shows the comparison of building layout and zoning of the two temples. It is found that the arrangement of Buddhavas and Sangkhawas zones of Wat Thour Thong and Wat Sam Lae are quite different. In Wat Thour Thong, the Buddhavas zone's layout formed in 'L' shape similar to the layout investigated from previous study [5]. Whereas Wat Sam Lae, the Buddhavas is surrounded by Sankhawas and formed 'L' shape layout arrangement. According to the field observation and interview from the abbot, the major strategy in planning the temple's layout and location emphasizes on the accessibility from the surrounding context. The study discovered that the layout is modified from its original setting. In the past, the main access to the temples was from the water-based access the Chao Phraya River but, at present, the temple's entrance is from the land-based access the main road.

(a) Wat Thour Thong

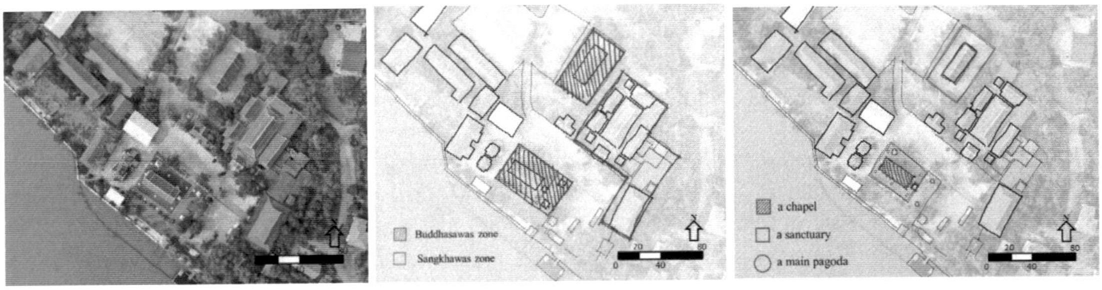

(b) Wat Sam Lae

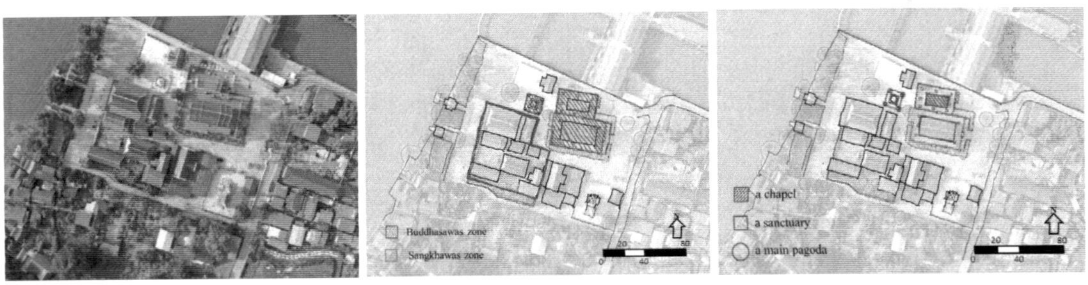

Figure 5. Comparison of building layout (middle) and zoning (right) Wat Thour Thong (a) and Wat Sam Lae (b).

4.2. Architectural design and stupa style

In Buddhavas zone, the arrangement of three architectural elements: stupa, chapel, and sanctuary, are not aligned according to the center of universe axis shown in Table 1. The locations of the stupa in these two temples are slightly shifted from the faithful Buddhist and universal paradigm founded in royal temples. It might be because these two temples were built by the communities. Comparing the scale of these three architectural elements with those of the royal temples, the size of stupa and sanctuary of these two temples are smaller due to the lower hierarchical significance of the temple. There is a significant difference in the style of the stupa design between Thai and Mon Buddhist temples, as presented in Figure 6. Mon stupa (b) is influenced by Burmese style, which has three lotus bases and it is surrounded by four small stupas.

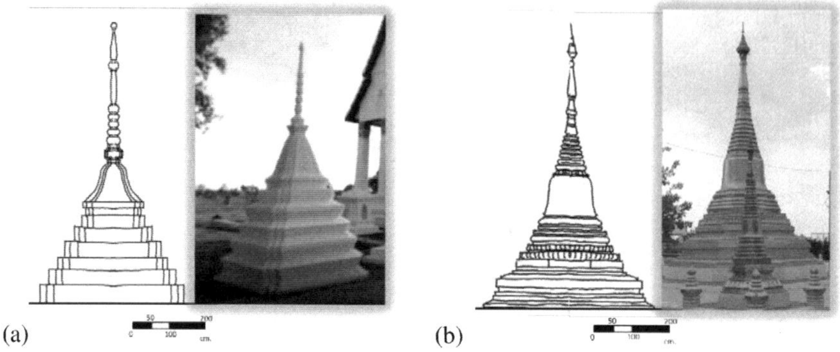

Figure 6. Architectural style of Thai Buddhist stupa (a) and Mon Buddhist stupa (b).

5. Conclusions

This study emphasizes on the temple's layout and architectural design influenced by Thai and Mon Buddhist communities. The study is conducted through field observation and interview with the abbot. The selected case studies of two temples, namely Wat Thour Thong represents the Thai Buddhist temple and Wat Sam Lae represents the Mon Buddhist temple. It is found that the Thai and Mon Buddhist temples have similar design concept in term of the arrangement of the stupa, the chapel, and the sanctuary in Buddhavas zone. The layout of Buddhavas and Sankhawas zones that their functions also shares similarity in zoning programs and its design according to the accessibility from the context. The architectural design and layout planning of these two temples are not aligned on the main axis, which differ from those of the royal temples. The major function of the community temples is to serve religious ceremonies resulted in the size of large chapel while the scale of stupa and sanctuary are smaller than those of the royal temples. However, the relationship of the layout planning of Buddhavas and Sangkhawas is the same as the royal temple arrangement. The only difference in the temple's character between the Thai and Mon Buddhist temples is in the stupa design. The Mon Buddhist temples is influenced by Burmese style. In the later stage of the research, the study plans to collect detailed data of the stupa design. Furthermore, more field observation will be conducted in more temples in order to prepare the database to support the cultural tourism development in Sam-Khok district. The limitation of this study lays upon the limited knowledge and experience in architecture of the key stakeholders. It is limiting the verbal references of this study, The careful planning to work around this limitation is necessary to achieve the more in-depth socio-cultural dimensions data and integrated analysis with the contextual summary of the site.

6. References

[1] Suvannin W. *A study of cultural tourism potentiality of buddhist templesin pathum thani province*. Journal of Graduate Studies Valaya Alongkorn Rajabhat University. 2014;8(3):78.
[2] Svasti P. Potty about pottery. Bangkok Post. 2016 17 November 2016; Sect. Life.
[3] Wiseesinth N. *The potential of buddhist tourism destination: temples in pathum thani province*. Journal of Graduate Studies Valaya Alongkorn Rajabhat University. 2017;11(3).
[4] James A. LaGro Jr. *Site analysis informing context-sensitive and sustainable site planning and design 3rd edition*. 2013.
[5] Jirathasanakul S. Wat: Buddha of Thai Architecture. 1, editor. Bangkok: Thammasat Printing House; *2001*.

Acknowledgement
This study was funded by the research fund from Thammasat University in 2018 (Grant No. 125/2561) and this research has been approved by a research ethics committee of Thammasat University.

IOP Conf. Series: Materials Science and Engineering **690** (2019) 012012 doi:10.1088/1757-899X/690/1/012012

The Origin and Development of Boxes in Sports Architecture

Qiuju Mao[1,2], **Siwen Guo**[1,3] **and Dawei Lin**[1,4]

[1] College of Architecture and Urban Planning, Tongji University, 1239 Siping Road, Yangpu District, Shanghai, China
[2] Guizhou Architectural Design Research Institute Co. 28 Lincheng West Road, Guanshanhu District, Guiyang, Guizhou, China
[3] Shenyang Jianzhu University, 9 Hunnan East Road, Hunnan New District, Shenyang, Liaoning, China
[4] Tongji Architectural Design (Group) Co., Ltd, 1230 Siping Road, Yangpu District, Shanghai, China

Email: 240738267@qq.com

Abstract. Since the 1990s, under the pressure of marketing, the box was set up in the NBA (National Basketball Association) arena in the United States, which made the application of the box in the stadium popular. The study of boxes is of far-reaching significance for stadiums and gymnasiums. It not only improves the current situation of poor operation and low utilization rate of stadiums and gymnasiums, but also responds to the needs of consumers for the commerciality and entertainment of stadiums and gymnasiums.

1. The definition of the box

The term "box" originally means a single-seat seat, which was initially found in architecture. The box of sports architecture first appeared in the Ancient Colosseum. The definition of the 'Box' here is: different from the ordinary open audience seats, usually a luxuriously decorated single room surrounded by four walls, with its own stand, allowing ten to thirty people to watch the game (figure 1) [1].

Figure 1. Inside the StockholmWukesong Basketball Hall

Figure 2. Roman gladeship box position

Figure 3. Location of the Stockholm Olympics

2. The origin of the box

In ancient Rome, special unit-style seats and spaces which we now call the box, appeared in sports architectures such as the Colosseum (figure 2). The Colosseum is an oval of 20,000 square meters. There are 60 rows of seats around the central performance area, which can accommodate 80,000 people. The seats with own stairs and access are divided into five zones in order to serve different classes. In the first row, there are marble boxes, especially for the emperors and officials. The box of the sports architecture as a manifestation of the authority of the official originated from here[2] [3].

3. The development of the box in the Olympic Stadium

From the first modern Olympic Games held in Athens, Greece in 1896, the Olympic architectures have attracted worldwide attention. The box, along with the development of Olympic stadiums, has been developed.

3.1. Appearance of the box in the modern Olympic architecture

The first appearance of the box in the modern Olympic architecture was in Sweden in 1912, the Stockholm Olympics. The Korolev Stadium designed by the Swedish architect Torben Grut has officially incorporated the design of the box into the sports architecture. In the original design, there were only 7,500 seats in the stadium. In the middle of the east seats are the royal seats. The seats below are converted into box styles. Each box can accommodate 4-10 people with better view and larger size than usual. (figure 3).

3.2. Development of Boxes in Modern Olympic Architectures

3.2.1. Stadium. With the development of the stadium, the number of boxes has gradually increased. Its position gradually moves from the top down to the middle of the two-story stand, which not only has the best sight, but also no longer affects the overall roof shape of the stadium(table 1).

Table 1. The development process of the box in famous Olympic Stadiums

Tokyo, 1964	Mexico, 1968	Munich, 1972	Moscow, 1980

Barcelona, 1986	Athens, 2004	Beijing, 2008	London, 2012

At first, all the boxes of the stadium have always been the place for leaders. Inspired and driven by the box of the professional gymnasium, the box of the Olympic Stadium has began to sell tickets. For example, in 2008 Beijing Olympic Stadium, the Bird's Nest, there are 140 boxes. And recently, one $60m^2$ box was rented at a price of 8 million for 30 years. Such huge profits have undoubtedly become the fuse for the rapid development of the box in recent years.

*3.2.2. Gymnasium.*The development of the box in the Olympic stadium has been very slow. On the one hand, the Olympic Stadiums are mostly funded by the government, they have not paid much attention to the operation of the venues. On the other hand, there are few competitions after the Olympics, which means the low attendance rate resulting in little profit of the boxes.

In recent years, the design of basketball and tennis courts in the Olympic Games is in line with the design of the professional league's arena. Most of the arenas will be handed over to the professional operation after the Olympics. The professional league and large-scale concerts are frequently held as a profit point. At this time, the number of boxes in the arena will have a quantum leap and the design will be improved.

For example, Basketball Hall of the 2010 Guangzhou Asian Games - Guangzhou Sports Performance Center (figure 4) is the first NBA standard basketball stadium in South China. The hall has designed 60 ring-shaped boxes, which are embedded in the design and business concept of the NBA Arena[4].

4. The development of the box in the NBA professional league venues

The West club operating model is different from the Chinese. Western clubs often host large-scale sports competitions all year round. NBA has a number of clubs, each year holding about 1300 world-class basketball games. The audience can reach more than 20 million, which makes a total of nearly one billion dollars in profits. Below we take the NBA basketball hall as an example to introduce the construction process of the box in the professional league venue.

4.1. The appearance of the box in the NBA venue

After the establishment of the NBA in 1946, the construction of the arena has experienced more than 60 years of history. Each club has a special arena as the home court. At the 1960-1980 stage, a large number of teams joined the NBA, especially in the 1976 NBA annexed the ABA (American Basketball Association), which laid the foundation for the absolute strength of the league in the US basketball. The expansion of the alliance has contributed to the renovation of the stadium. Almost all clubs have been renovating the stadium during this period. For example, the Houston Rockets, from 1975 to 2003, the team settled in the Compaq Center, using the stadium for 28 years. The venue has a total of approximately 16,200 seats and 20 boxes on the top floor (figure 5). This is the first time the box has appeared in the NBA venue[5].

Figure 4. Guangzhou Sports Performance Center in line with the professional league

Figure 5. Compaq Center Box Location

4.2. The development of the box in the NBA venue

Since the 1990s, NBA venues have witnessed a new trend of construction and reconstruction. In order to increase the utilization rate and profitability of the venue, most basketball halls have designed a large number of boxes to obtain high profits for tickets.

The reconstruction of NBA basketball hall is often because of the number of the boxes is too low rather than the scale of the hall. For example, the Golden State Warriors' stadium as one of the oldest stadiums in NBA was put into use in 1966. Before the start of the 1997-1998 season, the arena was completely refurbished with 72 luxury boxes added (figure 6) [6].

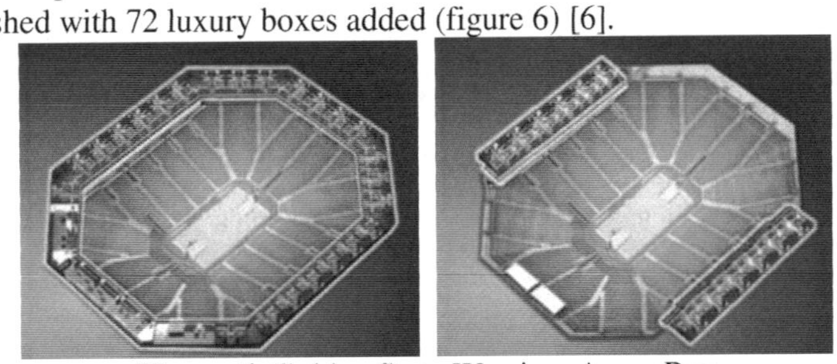

Figure 6. Golden State Warriors Arena Box

The following table is about representative stadiums in the NBA since 1990. We can see that the number of boxes in them is at least 50, and most of them have more than 90 (table 2).

Table 2. Box situation in the NBA arena

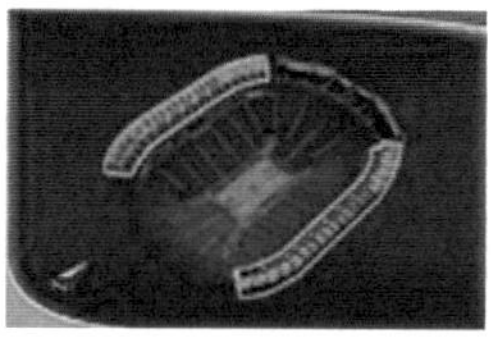		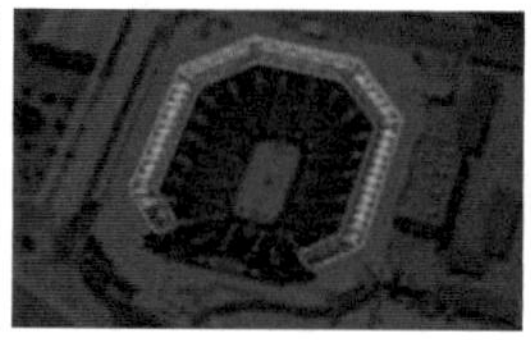	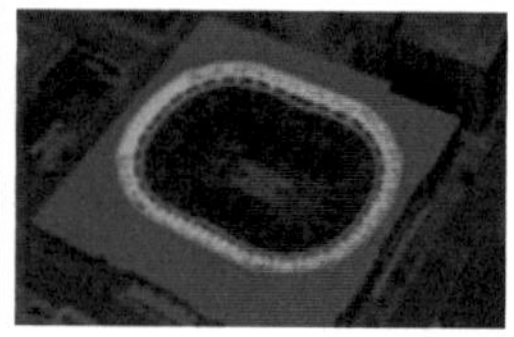
Houston Rockets Toyota Center Arena 103	Detroit Pistons Auburn Hills Palace Arena 180	Phoenix Suns Messi Arena 94	Washington Wizards MIC Center Arena 114
	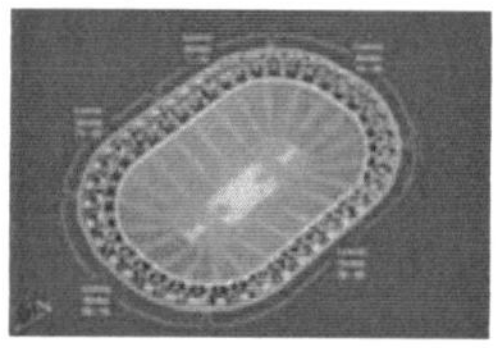	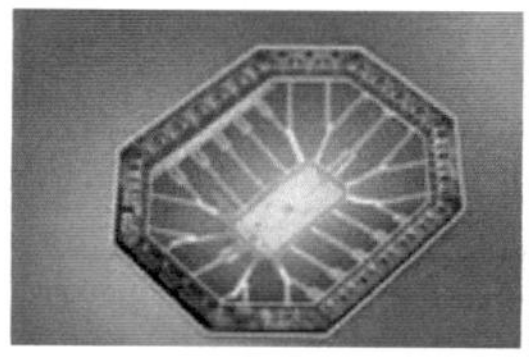	
Portland Trailblazer Rose Garden Arena 70	76 people in Wachovia Center Stadium 80	Golden State Warriors Auckland Stadium 70	New Orleans Hornets 53

5. The development of boxes in Chinese Sports Venues

From the 1990 Asian Games to the 2008 Olympic Games, and then to the 2010 Asian Games, the design of the box in the stadium has also gone through the process from scratch to becoming more and more rational in China. However, because of the institution, large stadiums will only be built by government when major events are held. There are no mature clubs to build stadiums themselves. Therefore, we are here to take the stadiums built by government as an example to see the development of the box.

In China, the number of boxes is increasing and the level of the design is being improved; but the box is only designed in basketball and tennis stadiums, which can hold the professional league (table 3) [7].

Table 3. Box development of venues in China's large-scale events

stadium	 Beijing Asian Games Workers Stadium 0 box	 Beijing Olympic Bird's Nest 140 boxes	 Guangzhou Asian Games main stadium 117 boxes
Basket-ball Hall	 Beijing Asian Games Glorious Stadium 0 boxes	 Beijing Olympic Wukesong Basketball Hall 45 boxes	 Guangzhou Asian Games Sports Performance Center 60 boxes

Descrip-tion	1990 Beijing Asian Games	2008 Beijing Olympic Games	2010 Guangzhou Asian Games
	During the 1990 Beijing Asian Games, facilities such as athletic field, swimming pool, gymnasium and hockey stadium were built. In the athletic field, technical special boxes such as radio and television are set up, and there is no box specially designed for the audience to watch the game. There is also no box design for the gymnasium or the swimming pool It can be seen that due to historical conditions at the time, China's sports architectures are still relatively backward compared with Western sports architectures at the same time.	Box appears during the Beijing Olympics. It is combined with the business, and its function is changed from a single to a composite. It also has a special streamline that is closely integrated with the VIP lounge. The box is more perfect in terms of area ratio, internal function setting and interior decoration. Wukesong Stadium is the venue for the Beijing 2008 Olympic basketball competition. There are 45 luxury boxes in the hall, 29 large boxes and 16 small boxes.	The main stadium of the Guangzhou Asian Games designed 117 boxes. As the main venue of the Asian Games, Guangzhou International Sports Performance Center has made a breakthrough in the design of the box. It is a new platform for world-class sports entertainment, integrating sports and performing arts activities. The venue adopted the commercialization model of the US NAB. With a total construction area of 120,000 square meters, the venue has 18,000 auditoriums and 60 360-degree luxury VIP boxes. And there is a private bar and club area behind the seats.

6. Conclusion

The development of boxes in Chinese venues still lags far behind those of Western countries. By comparing the differences of sports concepts between the West and the East, this paper introduces the development process of boxes in sports buildings in Western countries, and the development of boxes in stadiums and Gymnasiums in large-scale competitions in China in recent years.With the development of the globalization of sports and the importance attached by the owners to the later operation of sports buildings, the importance of the box will become increasingly prominent. I hope this article will help the design of future sports architectures [8].

References

[1] SUN Yuan, SUN Yifang and CHEN Yuanxin 2010 *J. Sports Culture Guide. Analysis of the Cases of Sino-US Sports Clubs and Their Enlightenment* **08** 71
[2] Feng Wei and Jie Yue 2009 *J. Urban architecture. Discussion on the design of modern sports building box facilities* **9**
[3] Zhang Wei'an, Su Tiexiang and Su Yi. 2010 *J Architectural creation. Guangzhou International Sports Performance Center* **11**
[4] Pan Yong and Chen Xiong. 2010 *J Architectural creation. Guangzhou Asian Games Museum design and thinking* **11**
[5] Guan Xuanhui and Luo Wei 2007 *Asian Games Basketball Competition Hall: Huangpu Sports Center Gymnasium* （Beijing: China Building Industry Press）
[6] Mei Jikui 1999 *Modern stadium architectural design* (Harbin: Heilongjiang Science and Technology Press)
[7] Qian Feng, Ren Lei, Chen Xiaoyu 2011 *Centennial Olympic construction* (Beijing: China Building Industry Press)
[8] Chi Jun 2007 *NBA Arena research* Master's thesis of South China University of Technology

IOP Conf. Series: Materials Science and Engineering **690** (2019) 012013 doi:10.1088/1757-899X/690/1/012013

Adaptation of Apartments to Small Hotels: The Case of Nakhon Si Thammarat Thailand

Chaiwat Riratanaphong[1]* and Chutima Kaewchan[1]

[1]Faculty of Architecture and Planning, Thammasat University, Pathumthani, Thailand

*Email: chaiwat@tu.ac.th

Abstract. The support of tourism industry from Thai government leads to the increasing demand and development of accommodation in the less visited cities including Nakhon Si Thammarat, which has been mostly visited in comparison to the other less visited cities since 2017. There is a high demand in small hotel accommodation type in Nakhon Si Thammarat whereas apartment buildings are oversupplied and becoming obsolete. This study aims to propose a step-by-step process for the adaptation of apartments to small hotels in Nakhon Si Thammarat, Thailand. Research method includes interviews with tourists, small hotel developers, the apartment developer and the expert in the adaptation of buildings to small hotels. The findings show the alignments between data from the interviews and what have been found in the literature that were developed to a proposed step-by-step process on the adaptation of apartments to small hotels. This study provides the adaptation process based on the case of Nakhon Si Thammarat, Thailand. More cases are required to test the step-by-step process in order to increase the validity and to make further progress. Further study could focus on feasibility analysis and the implementation of the step-by-step process in other cities to find out the similarities and dissimilarities regarding the different contexts of the cases.

1. Introduction

Thai government initiated several projects such as public utilities, main transportation routes and accommodation with the aim to improve local business in order to support tourism industry. With the increasing growth rate at 9.47 per cent from last year, Thailand's tourism industry generated income with the total number of 2.75 trillion Baht (90.55 billion USD) in 2017 that disseminated to every part of the country [1]. In 2016, data from the Tourism Authority of Thailand showed the increasing percentages of international tourists travelling to Thailand (2.8-4.8 per cent) and Thai tourists (6.2-7.2 per cent) travelling within the country. Government's plan to promote the less visited cities brought about the attention of investors and local communities to develop accommodation in the cities using tourism industry as a driver [2]. Nakhon Si Thammarat in Thailand is regarded as one of the most popular destinations among other less visited cities because of its' famous tourist attractions such as Phra Mahathat Woramahawihan Temple, Nakhon Si Thammarat National museum, Dohok Pavilion and Ho Phra Sung. Accommodation business in the city has gained benefits from the growth of tourism industry as a result.

2. Accommodation's demand and supply in Nakhon Si Thammarat

The development of accommodation strategy of Nakhon Si Thammarat city resulted in the increasing number of tourists and a higher demand in accommodations whereas accommodation supply has been fairly limited. The total number of 132 accommodation units in Nakhon Si Thammarat comprise of

60 apartments (45.45 per cent), 38 hotels (28.80 per cent), 17 dormitories (12.90 per cent), 15 mansions (11.35 per cent), 1 guest house and 1 resort (0.75 per cent each).

2.1. Apartment
Apartment is a type of accommodation that provides individual dwelling for rent for more than 4 units. Each unit includes bathroom, toilet and kitchen and may provide a multi-purpose space or include 2-4 bedrooms. The unit is provided with standard furniture such as cabinets, beds, air conditioners, fridges, water boilers, televisions or without furniture. The building is designed for renters to walk up the stair without the provision of an elevator. Usually, the floor space per unit can be more than 100 square metre depending on building grade or price per unit of each apartment. Renters have to comply with apartment's rules and with specific occupying period [3]. Renters include students, families without children, young couples or company workers that have to move from place to place.

2.2. Small hotel
Hotel is set up with a business objective in providing temporary accommodation to visitors. It provides paid lodging on a short-term basis. Small hotel is defined as the hotel that has no more than 30 rooms. In this study, small hotels are selected from cases with room availability ranging from 3-30 individual units and have more than 20 visitors [4]. There are four main functions of small hotels including:

- residential - individual unit provided as: 1) private room with bathroom, 2) dormitory type with sharing bathroom. The space allocation in each unit can be classified on different hotel grades: budget/economy, mid-price, upscale and luxury.
- front of the house - includes reception, common area, dining area and multipurpose space.
- back of the house - includes functions used by hotel staff such as storage, machine room, staff room, lockers, personnel office and kitchen.
- service space - includes parking, surrounding area, circulation and stair hall.

3. Adaptive reuse
Adaptive reuse is defined as the renovation and reuse of pre-existing structures for new purposes [5]. In this study, the term refers to the adaptation of old building apartments to small hotels with regard to the requirements of visitors. Findings from the literature show drivers of adaptive reuse that include the potentials to: 1) create usability of the new function, 2) improve quality of the building, 3) reduce acquisition time in comparison to new building construction, 4) provide investment advantages. Three main physical condition factors that affect the implementation of adaptive reuse are: 1) structure of the building, 2) construction technique, 3) building characteristics.

4. Problem formulation and research question
There has been an increasing demand of accommodation in Nakhon Si Thammarat with regard to the government policy in the support of tourism in less visited cities. Apartments as accommodation supply are currently provided to tourists in the area, but do not respond to their needs whereas small hotels are becoming more popular. The purpose of this study is to propose a step-by-step process for the adaptation of apartments to small hotels in Nakhon Si Thammarat. The paper aims to answer the following question:

- How can adaptive reuse concept be applied to apartment buildings to solve the increasing accommodation demand in small hotels?

5. Research method
Data collection of this study is through interviews to obtain qualitative data and in-depth information. Four groups of participants' interviews include: tourists (5 persons), small hotel developers (5 persons), the apartment developer and the expert in the adaptation of buildings to small hotels.

- Interview with tourists aims to collect data on preferences and the decision to stay in small hotels.

- Interview with small hotel developers focuses on building design and factors influencing business operation of small hotels.
- Interview with the apartment developer collects data on business operation, building information, occupancy rate, target customer and duration of stay.
- Interview with the expert in building adaptation focuses on the steps towards the adaptation of apartments to small hotels.

6. Findings

The findings include four parts: 1) interview with tourists, 2) interview with small hotel developers, 3) interview with the apartment developer and 4) interview with the expert in the adaptation of buildings to small hotels.

6.1. Interview with tourists

Findings from the interviews with tourists travelling to Nakhon Si Thammarat during annual festivals of the city are described as follows:

6.1.1. General information of the tourists. The findings show that the majority of tourists are female from 41-50 years old with the average income between 20,000 - 30,000 Baht (651 - 977 USD) per month. Most of the tourists reside in the southern cities of Thailand and work as civil servants and company employees.

6.1.2. Tourist behaviours. Tourists visit Nakhon Si Thammarat choose to stay in small hotels because of the lower room price per night. In addition, the variety of hotels that offer range of services in various locations make it more convenient for tourists in terms of accessibility. Generally, tourists visit the city in a group of 3-4 persons. The common destinations are cultural and historic places followed by religious sites. The findings show the average duration of stay of 3 days and 2 nights. Surprisingly, tourists do not search for accommodations through the internet in advance, but choose to ask for information directly from the hotels.

6.1.3. Factors influencing accommodation selection. The findings show factors that influence small hotel selection including:

- location - hotels situated close to tourist destinations within walking distance or within the travelling by public transport gain advantages in terms of accessibility.
- services and sales promotion - includes services from personnel staff such as personality and accessibility and the ability to provide sufficient information to customers. Sales promotion such as the offer of morning meal or local food/beverage as a complimentary from the hotels draws tourists' attention.
- interior and exterior of the building - interior and exterior of the building with distinct characteristics and identities promotes attractions.
- room price - findings from the interviews show that the average room price per night are from 1,000 - 1,500 Baht (31.43 - 47.14 USD) per night. The tourists are willing to pay in higher price if the conditions of rooms and facilities as well as hotel services are properly provided.

6.2. Interviews with small hotel developers

Findings from small hotel developers regarding factors influencing the development of small hotel projects in Nakhon Si Thammarat are described as follows:

- location and accessibility - this item is considered as the most important factor. The hotels must be located on main roads and easily accessible from public transport and close to tourist attractions. This item aligns with requirements of the tourists.
- exterior design and interior decoration - the hotels should express distinct characteristics of the city such as cultures, traditions, materials and local climates both in terms of architectural

design and interior decoration. This item corresponds with findings from interviews with the tourists.

- supporting facilities - usually, most hotels provide standard facilities such as parking, smoking area, shower room and dining area. However, there is increasing demand of facilities for elderly and disabled visitors that should be provided.
- amount of rooms - the findings show the average number of 20 rooms for small hotels with the average price per night of 885 Baht (28.90 USD).
- size of rooms - the findings show the average room area of 28.50 square metre.
- usable area in the building - room area has the highest percentage of the total usable area (40 per cent) followed by service area, i.e. walkways, parking spaces, stair halls (28 per cent), front of the house such as reception, dining area, shops (24 per cent) and back of the house (8 per cent).
- After sale service and promotion - this item includes daily cleaning, luggage storage service, transportation service and travel service offered by hotels.

6.3. Interview with the apartment developer

Findings from the interview with the apartment developer in Nakhon Si Thammarat provide information on physical characteristics of the building, project marketing and finance as well as challenges of apartment business that can be described as follows:

- physical characteristics - buildings are located in secondary road connected to main road and are close to tourist attractions. With only one type of room offered for the apartment, the total number of 18 rooms are provided with bathroom and fan.
- the average rent is 2,300 Baht (74.71 USD) per month with the required deposit up front. The average duration of stay is 5 months. The majority of renters are civil servants.
- challenges of apartment business - the major problem is the inconsistency on due payment or left without pay from renters followed by quarrels between renters and physical obsolescence of the buildings.

6.4. Interview with the expert in the adaptation of buildings to small hotels

Findings from the interview with the expert regarding the adaptation of apartments to small hotels are adapted in connection to what have been found in the literature and proposed as a step-by-step process as follows table 1:

Table 1. A step-by-step process for the adaptation of apartments to small hotels

Steps	Details
1	Examining laws and regulations - examine urban planning laws, i.e. examine building location whether it can be applied for hotel business - examine regulation on building transformation and adaption, i.e. examine building location associated with regulation on building transformation and adaptation to hotels
2	Performing preliminary feasibility analysis - analyse location potential - set up project initiation plan - study physical characteristics - conduct financial and marketing analysis
3	Surveying surrounding areas and building layout - collect data on surrounding area - collect data on existing building, i.e. examine building conditions and consider which parts have to be renovated or put in current condition.
4	Setting up business plan - plan for business operation - plan for building design operation

Steps	Details
5	Developing architectural design of the project - develop architectural design according to laws and regulations on building adaptation to small hotels
6	Authorizing building adaptation - inspect building to be aligned with building laws and regulations - request for permission on building adaptation
7	Conducting building adaptation - conduct building adaptation from structures, architecture works, surroundings, building systems, facades to interior spaces - request for permission on the changing of building function
8	Submitting permission to operate hotel business
9	Preparing for building operation - inspect building functions together with architects and engineers - request legal documents from government authorities - prepare building operation plan - practice the use of building functions - prepare hotel management system (e.g. reservation and accounting system) - prepare spare parts of building in case of long term maintenance - plan for marketing and sales system

7. Conclusions

The study shows factors influencing the adaptation of apartments to small hotels in Nakhon Si Thammarat. The most important factor is location that must be easily accessible from public transport and close to tourist attractions. Regarding the physical characteristics of buildings, the adaptation on size of rooms and walkways and fire escape have to rely on laws and regulations of the area. Interior and exterior design of the building need to follow cultures, traditions and location climates. The other functions in common areas such as reception, dining area and shops have to be properly arranged. The provision of related services such as transportation and travel service is important to the decision to stay of the tourists. A step-by-step process for the adaptation of apartments to small hotels includes: 1) examining laws and regulations, 2) performing preliminary feasibility analysis, 3) surveying surrounding areas and building layout, 4) setting up business plan, 5) developing architectural design of the project, 6) authorizing building adaptation, 7) conducting building adaptation, 8) submitting permission to operate hotel business and 9) preparing for building operation.

There are certain alignments between findings from the interviews with different groups of participants. Data from the interviews with various stakeholders involved in the adaptive reuse project is useful for developing the step-by-step process. This study provides the adaptation process based on the case of Nakhon Si Thammarat, Thailand. More case study buildings are required to test the step-by-step process in order to increase the validity and to make further progress. Further study could focus on feasibility analysis and the implementation of the step-by-step process in other cities to find out the similarities and dissimilarities regarding the different contexts of the cases.

8. References

[1] Economic Tourism and Sports Division. (2017). *Tourism Economic Review*. Bangkok Thailand: Ministry of Tourism and Sports.

[2] Thai Post newspaper. (2018, May 2). Government supports less visited area to promote country's economy. *Thai Post newspaper*. Retrieved from https://www.thaipost.net/main/detail/34337

[3] Pimsiripanich, S. (2014). *Factors Influencing the Selection of Apartments for Rent in Nava-Nakorn Industrial Estate After the Floods in 2554 B.E.* (Master of Science), Thammasat University, Pathumthani.

[4] Pornprathanwech, T. (2015). *Development of Small Hotels for Promoting Cultural Tourism: A Case Study of Renovating Shophouse in The Urban District of Ratchaburi Province.* (Master of Science), Thammasat University, Pathumthani.
[5] Adaptive reuse. (n.d.). In *Merriam-Webster's online dictionary* (11th ed.). Retrieved from: https://www.merriam-webster.com/dictionary/adaptive%20reuse.

Research on Evaluation Index of Energy Efficiency Design of Multi-level Sports Complex Based on Synergistic Effect

Siwen Guo[1,2], Qiuju Mao[1,3] and Jiamin Zou[1]

[1] College of Architecture and Urban Planning, Tongji University, 1239 Siping Road, Yangpu District, Shanghai, China
[2] Shenyang Jianzhu University, No.9, Hunnan East Road, Hunnan New District, Shenyang, Liaoning, .China
[3] Guizhou Architectural Design Research Institute Co. 28 Lincheng West Road, Guanshanhu District, Guiyang, Guizhou, China

Email: g_siwen@163.com

Abstract. Transforming the single-storey large space of traditional sports buildings into vertical extension space of multi-level venues, the multi-level sports complex achieves efficient use of resources by introducing a combination of recreation and social functions. Analyzing the key design points of the cascade-type gymnasium based on relevant criteria and standards of evaluating sports buildings, this paper establishes a multi-layer evaluation system to achieve synergistic energy efficiency design. It also provides a reference for expanding the efficient design of sports complexes.

1. Introduction

By 2013, the total number of sports venues per ten thousand people in China has nearly doubled vs 2003. In 2015, the number of sports venues in China was approximately 1.88 million, and the national stadium area reached 2.53 billion square meters. Large stadiums and gymnasiums, as essential building types, possess three characteristics: large initial investment, high operating costs and large energy consumption. The multi-level sports complex can greatly save enormous space and alleviate the pressure of high-density environment in city.

2. Multi-level sports complex

Superimposing multiple sports (site) spaces and other related functional spaces in the vertical direction, the multi-level sports complex takes advantage of large-span construction technology to transform the single large space of traditional sports buildings into multi-layer composites. For example, Zhonghe Sports Center, located in New Taipei City, provides public facilities for a range of sports activities including badminton, basketball, rock climbing and hockey. "Lacking spatial and visual qualities is often the aftermath or syndromes of this building type, particularly in the metropolitan cities of Taiwan," said the architect. (figure 1)

Figure 1. Section of Zhonghe Sports Center.

In the first half of the 20th century, developed countries such as European countries, America, and Japan began to study large-span buildings with overlapped multi-layer space. From the perspectives of urban development, economic planning, and sports space, they explored multi-storey large spaces and began to design small community multi-storey gymnasiums with commercial functions. Furthermore, there are more than 10 thesis studying in training halls, college gymnasiums, and hybrid structures of multi-level large-span building. A comparatively new research, *From single to compound: research on the design of multilevel gymnasium based on national fitness,* by Professor LU in 2018 from HIT China, analyzes the key design points of the cascade-type gymnasium based on the in-depth study of the function and technology of the laminated spaces[1]. Skyworld Arena, a 3.80ha sports complex, just completed in Malaysia, is a good reference. There are also many evaluation criteria of sports building.(Table 1)

Table 1. Relevant evaluation criteria of sports building.

Standard	Standard number	Country
Design Code for Sports Building	JGJ 31-2003	China
Specification for acoustical design and measurement of gymnasium and stadium	JGJ/T 131-2012	China
Standard for lighting design and test of sports venues	JGJ 153-2016	China
Technical specification for intelligent system engineering of sports building	JGJ/T 179-2009	China
Code for electrical design of sports buildings	JGJ 354-2014	China
Standard Specification for Helmets Used for Recreational Snow Sports	ASTM F2040-2018	USA
Developing Sustainable Sports Facilities. A Toolkit for the Development of a Sustainable Community Sports Hub	OHSIS ZBH 368-2008	USA
Sports lighting-Part: Daylighting	DIN 67526-3-2018	Germany
Sports halls. Halls and rooms for sports and multi-purpose use. Part 1: Planning principles	DIN 18032-1-2014	Germany
Sports halls-Martial arts halls-Design, layout and use	NF P90-209-2016	France

The existing research usually limits its discussion to the concepts around feasibility, such as space combination and technical difficulties, while systematic research on the energy efficiency design of the cascading sports complex from the view of synergistic effect is limited.

3. Construction of energy efficiency evaluation system for multi-level sports complex

3.1 Energy efficiency design based on synergistic effect

Synergetics is to study various systems, composed of a large number of subsystems under certain conditions, through the synergy between subsystems, in a macro ordered state, form a self-organizing structure mechanism with certain functions. The system evolution mechanism is called order parameter. For this research, the functional structure construction principle, the overall goal, the combined effect, and the functional compatibility play the role of order parameters, thus promoting the optimization of the system.

Sports complexes should apply appropriate ecological technologies under environmentally sustainable development design concepts. Since the multi-level sports complex also consumes a lot of energy, it is helpful to simplify the energy efficiency index and thus test the "greenness" under different schemes by considering the physical performance of different sub-venues and the energy efficiency of the structural system.[2]

3.2 Establish the evaluation model

The energy efficiency design evaluation index model of the multi-level sports complex is divided into the target layer, the criterion layer and the index layer. The target layer refers to the evaluation of the multi-level sports complex in terms of synergetics energy efficiency design. The criterion layer classifies each element according to relevant sports building codes and standards, and has guidance. The indexes explain the index layer into detailed design elements [3][4][5].

3.3 Clarify the indexes

3.3.1. Physical performance. In a multi-level venue, the thermal comfort requirements of different types are different and affect each other. Factors such as wind, light and thermal environment of different sub venues need to be considered and adapted separately in order to satisfy the comfort standards of each venue with high efficiency . The adjacent venues should avoid the heating loss through the partition or floor, and the public space connection between the venues can serve as an effective physical environment buffer or barrier. At the same time, it is necessary to control the equipments of each site separately, and fully consider the maximum load of the environment and reasonably predicting the usage mode.[6]

3.3.2 Functional layout. The multi-level sports complex is more diversified in terms of functional layout than traditional sports buildings. The scale of the building, the number of building layers, the type and size of sports venues, and the height requirements should be considered when forming a reasonable layout. Analyzing the internal and external constraints of the functional structure enables the designer to understand the ways in which different parts of the building should be combined, and their respective suitable scale. This includes the layout of different levels and the number of indoor venues. In addition to providing sports-related functions, sports complexes often include commercial facilities such as sports training, sports lottery sales, monopoly of sports goods sale, and sports-themed catering. They mutually reach a rational combination to form a complementary and efficient use of the space[7].

3.3.3 Structure system. The stacked large-span structure system shares common characteristics with multi-level and long-span structures. Its stress performance under stress is similar to these two structures but more complicated. The types of structure should be selected according to different functions and spaces. Emphasis ought to be placed on the performances, such as the seismic performance and the ability to resist continuous collapse in the event of accidental loads, as well as special studies on the comfort analysis of buildings. The cascading structure system is generally a superposition of large space structure systems[8]. It is necessary to conduct a research from the perspective of economic cost and functional fit to form a structure system database and provide technical support for the application of the sports complex[9]. (Table 2)

Table 2. Evaluation index table of energy efficiency design of multi-level sports complex.

Target layer	Criterion layer	Index layer	Index connotation
Energy efficiency design of multi-level sports complex based on synergistic effect	Physical performance (U1)	Multi-venue lighting environment (V1-1)	Daylight, lighting
		Multi-venue wind environment(V1-2)	Ventilation, mechanical air supply, wind control
		Multi-venue thermal environment(V1-3)	Thermal comfort, humidity, temperature
		Public space transition(V1-4)	Heat transfer buffer, sound insulation
		Surrounding ecological environment(V1-5)	Site microclimate
	Functional layout(U2)	Construction scale (V2-1)	Regional population, frequency of use, openness
		Venue type and size(V2-2)	Intensive design
		Sports function layout(V2-3)	Venue layout
		Other function layout(V2-4)	Multifunctional operation of ancillary space
	Structure system(U3)	Rational and advanced structure(V3-1)	Seismic performance, resist continuous collapse
		Economical material(V3-2)	Recycled materials

3.4 Establish the weight of indexes

The indexes of designing multi-level sports complex are sophisticated and cover a wide range of parameters. The energy efficiency synergetic design of the complex is also a system involving multi-objective decision making problems. The indexes of the complex design a set of qualitative preparation plans, which assign and weight the index of the plan layer to converts the original divergent subjective qualitative judgment into a quantitative judgment matrix. This establishes a "qualitative-quantitative-qualitative" connection between the final design target and the design elements. By using the Delph method to compare the indexes, starting from the criterion layer of the hierarchical model, this research compares the different influencing factors of the hierarchy with the importance of the basic criteria in the previous level. The scale method is used to numerically identify the comparison results, quantifies the importance, in order to get the relative importance between each two indexes, then construct the judgment matrix. According to the formed weight index system, the synergy effect of energy efficiency design for the same target can be achieved [9].

4. Conclusions

The multi-level sports complex is growing larger and more complex, shares many similar attributes with sports buildings and urban complexes. Based on the design method of sports buildings, this research obtains the design index and weight database by using the analytic hierarchy process to connect the relationship between the design strategy layer and the criterion layer. It enables people to evolve from competitive functions to fitness functions, and even to social functions, thereby accelerating the construction of a shared commercial sports platform.

5. References

[1] Shiliang Lu, *From single to compound: research on the design of multilevel gymnasium based on national fitness,* Urbanism and Architecture, Volume 13, 2017, pp112-115
[2] Jonathan Mallie, *Expansive Workflows: Downstream Coordination in the Design of Sporting Facilities*, Architectural Design, Volume 87, 2017, pp 68-73
[3] M.Turrin, E.van den Ham, A.Kilian, S.Sariyildiz, *Integrated design of a large span roof: a parametric investigation on structural morphology, thermal comfort and daylight*, Proceeding of ICCBE 2010, International Conference, Nottingham UK,2010
[4] Ministry of Construction, PRC. Industry Criteria JGJ/T 131-2012: Specification for acoustical design and measurement of gymnasium and stadium(2012)
[5] Ministry of Construction, PRC. Industry Criteria JGJ 153-2016: Standard for lighting design and test of sports venues(2016)
[6] Ministry of Construction, PRC. Industry Criteria JGJ/T 179-2009: Technical specification for intelligent system engineering of sports building(2009)
[7] Ministry of Construction, PRC. Industry Criteria JGJ 354-2014: Code for electrical design of sports buildings(2014)
[8] Heiselberg P, Brohus H, Hesselholt A, et al. *Application of sensitivity analysis in design of sustainable buildings*. Renewable Energy, 2009, 34: 2030-2036.
[9] Green Olympic Building Research Group, *Assessment System for Green Building of Beijing Olympic*, China Architecture& Building Press, 2003:73-76

IOP Conf. Series: Materials Science and Engineering **690** (2019) 012015 doi:10.1088/1757-899X/690/1/012015

Analyzing Organization-to-project Interfaces in the Integration Management of Engineering Projects

Jialu Yu, Qing Yang* and Na Yang

School of Economics and Management, University of Science and Technology Beijing, No.30 Xueyuan Road, Haidian District, Beijing 100083, China

Email: yjlustb@hotmail.com

Abstract. To carry out the construction project management work effectively, the organization interface management in the workflow is a fundamental challenge.In the ongoing project, multiple project teams need to cooperate multi-party andcomplete project product delivery in different work interfaces. Organizational interface withinworkflow will help system planning and management in the engineering field. In this study, we first propose the evaluation indicators of engineering participants and tie strength between different organizations usingthe design structure matrix (DSM). Then we construct a multi-domain matrix (MDM) based on organization-to-project by correlating multiple participants in the project with the workload completion degree and present a comprehensive strength between the organization and the project to identify the crucial project interface.Finally, we give an engineering case to illustrate the proposed method and discuss its practical value. The proposed methods reinforce several managerial practicesfor the construction organization and project arrangement.

1. Introduction

In recent years, mega engineering projects have achieved great development and enhance the economic development. In complex construction projects, the participants face many risks such as high cost, long construction period and complex technology et al. To shorten the project construction period, reduce construction costs, and ensure the quality of construction production [1], it is essential to build systematic multi-domain linkage among project participants. Traditional project planning strategies and project management instrument are often inefficient [2]. Organization's competitive capability has received significant attention in engineering management [3]. To solve these problems of mega projects, we need an interaction interface management system that is capable of identifying and managing the organizations-to-project in phase of the project lifecycle.

A project can be viewed as an assemblage of organizations, work, information, and other attributes that have to be integrated to achieve project objectives [4]. Today, projects have become extremely complex with overlapping activities in the engineering construction project and certain time intervals among the deliverables of different project organizations (i.e. teams). Thus, project managers must deal with interfaces of organizations to projects than ever before. This paper focuses on collaborative relationship of participants and the subprojects with different degrees of interaction.

2. Literature Review

A crucial factor of successful projects is how to establish an appropriate organizational interface among organizational teams [5].An efficient organizational interface may reduce management complexity during the project process [2]. However, due to the complexity of team's interdependence, construction project organization architecture may not be properly established [6].Delivering of projects on time is significant for construction management [4].Interface management is used in many large construction projects to enhance coherence among stakeholders and reduce potential conflicts in projects [3].The researchers propose to combine process management with interface theory, but they do not construct participants interface management from the perspective of organization[7]. The Design Structure Matrix (DSM) method is a powerful network modeling tool to represent the elements involved in the interface and their interactions [5]. The Multi-Domain Matrix (MDM) is an extension of DSM modeling, in which two or more DSM models in different domains are represented simultaneously.

3. Proposed methodology

This paper proposed a methodology to illustrate the interfaces from organization to project. Organizations in the project need to exchange the elements of a task or a construction milestones with the project interface, which includes machinery, materials, facilities and processes. In this paper, DSM methods are used to measure the organizational structure, and MDM is proposed to analyze the dependence intensity between projects from organization to project in engineering projects [4].

As shown in figure 1, the steps of the model are as follows: First, analyse the mutual influence of the project participants from the perspective of investors' demand. Second, we set up an organizations' contribution (i.e. Percentage of project completion by each organization) as mapping matrix. Finally, the influence of different participants on the sub-projects in the construction project management is determined.

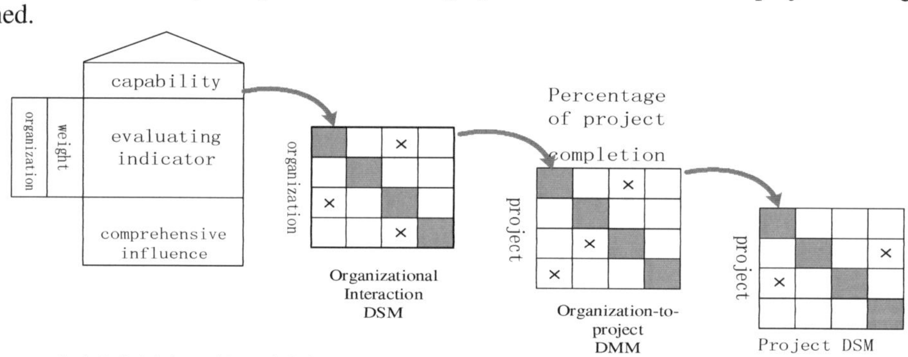

Figure 1.A Multi-level Model for DSM and analysis of dependency relationships between organizations and projects

3.1. Organizational interface capability evaluation

The evaluation for the competitive capability of an organization (or teams) in an engineering project is usually completed by the investor. In this paper, we propose "the measurement of organizational reputation level, team performance ability, resource management level, risk control ability, cost control level, green construction technology and project manager ability" as the evaluation criteria. The Organization interaction DSM is established as shown in figure 2.

		percents	Evaluating indicator						
			P_1	P_2	P_3	P_4	P_5	P_6	P_7
Organizations	*I*	0.3	9+	1−	5+			5−	1+
	II	0.3	1−	9+		5−	1+	1+	
	III	0.4		1+		5+	5+		5−
	results		3	3.4	1.5	3.5	2.3	1.8	2.3

(a)

Interaction

	I	*II*	*III*
I			
II			
III			

(b)

Figure 2. Organization measurement matrix and the Organization tie strength DSM

Next, evaluate the dependence of each index in the organization. This paper uses three degrees 1, 5, 9 to determine the organization interaction strength. The bigger the evaluation factor (i, k) is, the higher affection this parameter made. No number indicates that the evaluation indicator of organization will not affect the interaction among teams. It is noted that the"-"represents negative correlation of the evaluation index. After the weight and relation matrix of each criterion is determined, the relative importance value ω_k of each criterion can be obtained by Equation (1).

$$\omega_k = \sum_{i=1}^{m}(F_i \times OMM(i,k)) \quad i = 1,2,...,m, \quad k = 1,2,...,K. \tag{1}$$

In Equation (1), *Fi* indicates the weight, *OMM (i, k)* represents the value of row *i* and *k* column in OMM, which represents the affection between organization *I* and the competitive capability by evaluation index, *m* is the number of teams, and *K* is the number of evaluation parameters.
Based on above description and discussion, the organizational interaction matrix O_DSM (i, j) can be calculated by Equation (2) [7].

$$O\_DSM(i,j) = \sum_{k=1}^{K}(\omega_k \times I_{j\to i} \times OMM(i,k)) \times \left(\sum_{k=1}^{K}(\omega_k \times OMM(i,k))\right)^{-1} \tag{2}$$

where K is the number of evaluation indicators, $I_{j\to i}$ indicates the effect of the evaluation criteria on the organization, $I_{j\to i}=1$. The denominator represents the overall degree of correlation between the organizations *i* and all the evaluation indicators, and the numerator indicates the comprehensive influence of the relationship between the evaluation parameters.

3.2. Project DSM based on organization-to-project MDM

3.2.1. The interface of engineering project. The investor of the project will sign multiple engineering contracts with a number of project participants whom complying with an agreementin accordance with the responsibilities. Figure 3 shows that milestone delivery of building products occurs at different stages in different tasks. All the participants in the current task have formed the engineering project interface. D (i)represents of upstream activity i, D (j) represents downstream activity j, and the teams between the upstream and downstream exists in the different activities between i and j. Fig.3 (a) indicates that the participant will do the downstream activity after the preliminary activity is completed. Fig.3 (b) shows that the participant will do the upstream activity after the preliminary activity is completed.

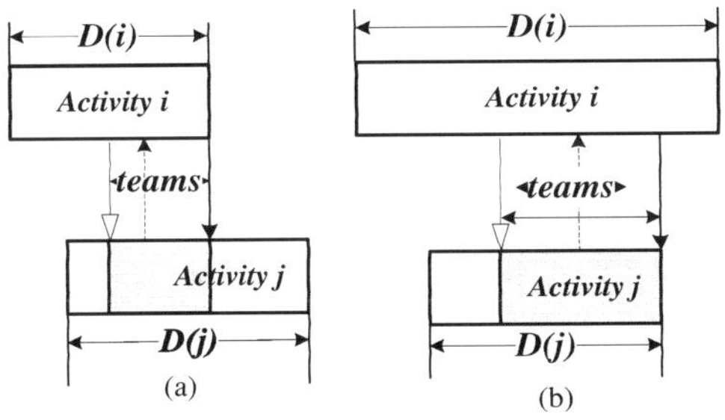

Figure 3. Construction milestones delivery

The coordination between the organizations is not only for a single task, but also for the progress of multiple tasks. Multiple organizations complete part of the subproject. Therefore, the intensity of interaction in organizational management highly depends on the coordination between the organizational units performing tasks.

We use the organization's contribution (e.g. percentage of project completion by each organization) to build the organization-to- project DSM. As shown in the figure 4 below:

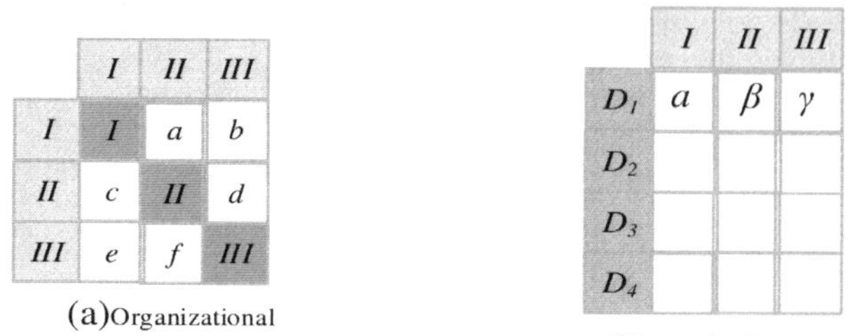

(a)Organizational interaction DSM

(b)organization-to - project DMM

Figure 4.Organization's contribution

The letters in Fig.5. (a) are calculated by the Equation (2),Fig.5.(b) is a typically non-square matrix mapping the domain of organizational interaction DSM to the domain of project DSM, which representsorganization's contributionin different engineering projects. It is important to note that the sum of the parameters α,β,γ is 1.

3.2.2. Organization-to-project MDM. The Multi-Domain Matrix (MDM) model can be utilized to reveal the relationship between different domains. MDM is an extension of DSM modeling in which two or more DSM models in different domains are represented simultaneously [1].As shown in Fig.7 (a) and (b), the MDM includes organization DSM (as shown in 3.1), project DSM and DMM (as shown in 3.2.1).

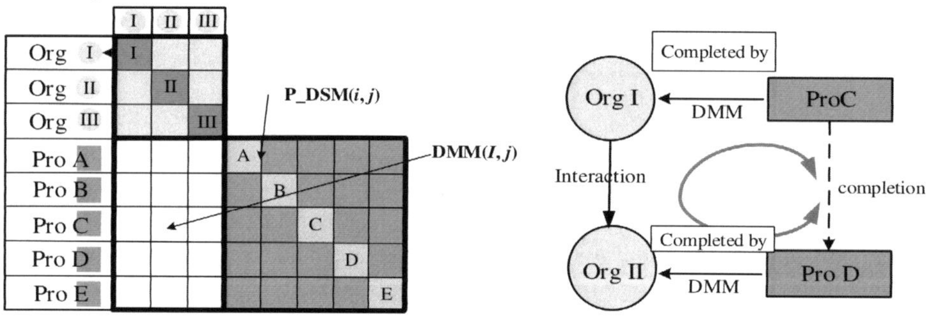

(a)Organizational Interaction DSM and project DSM based on

(b)cooperation by orgs between pros

Figure 5. Deriving project DSM between projects based on MDM

The dependency strength P_DSM(I,J) among projects (see Fig.6.(a)) which results from organizational relationships can be calculated with Equation (3)[7].

$$P\_DSM(I,J)=\sum_{i=1}^{m}(DMM(I,i)\times\sum_{j=1}^{m}(O\_DSM(i,j)\times DMM^{T}(j,J))) \quad (3)$$

where O_DSM(I,J) is the value of dependency strength between teams i and j in the org DSM, each row and column is an organization. DMM(*i,j*) is the mapping value between organization and project where its value indicates the organization's contribution in different engineering projects. P_DSM(*I,J*) represents the comprehensive intensity which are influenced by the capabilityof interface from the organizational interface.

4. An illustrative example

In this paper, we applied the proposed concepts and models to an example, a Steam turbine installation of Power plant project.According to the industry specification and contract division, the steam turbine installation project is mainly divided into the following five steps: A= The foundation of the turbine hall above the floor zero meter; B= The delivery of the turbonator's foundation; C=The sealing and the roof's end construction of the turbine hall's; D= steam turbine's platen in position; E= ST box up.

Through the reputation level of organization, organizational performance ability, resource management level, risk control ability, cost control level, green construction technology and project manager ability as the evaluation criteria. As shown in figure 6, the organizational evaluation matrix and the Organization interaction DSM are established.

		Per	Organization (team) evaluation index						
			P_1	P_2	P_3	P_4	P_5	P_6	P_7
Org	*I*	0.3	9+	1-	5+			5-	1+
	II	0.3	1-	9+		5-	1+	1+	
	III	0.4		1+		5+	5+		5-
	results		3	3.4	1.5	3.5	2.3	1.8	2.3

(a)

Org-interaction

	I	*II*	*III*
I		0.8	0.24
II	0.65		0.33
III	0.13	0.4	

(b)

Figure 6. Organizational evaluation and organization interaction DSM

Further, the organization interaction DSM is mapped into the "organization-project" matrix, as shown in Fig.7. It can be seen that the influence degree of project A to other projects is 0.78+0.7+0.73+0.9=3.11, and the influence degree by other projects is 0.7+0.74+0.68+0.92=3.04. Therefore, the leading role of project A in the entire project interface is 3.11/3.04=1.3; similarly, the leading roles of project B, C, D, and E in the entire project interface are: 0.67, 0.85, 0.71, and 0.8. Project B is mostly affected by other projects. Therefore, it is necessary to combine the interactions among organizational interfaces with the actual situation of projects to manage multiple construction projects.

O_DSM(*i*,*j*)	I	II	III					
Team I	I	0.8	0.24					
Team II	0.65	II	0.33					
Team III	0.13	0.4	III					
Pro A	0.3	0.4	0.3	A	0.70	0.74	0.68	0.92
Pro B	0.5	0.2	0.3	0.78	B	0.87	0.61	
Pro C	0.2	0.5	0.3	0.7	0.74	C	0.71	0.88
Pro D	0.4	0.2	0.4	0.73	0.58	0.81	D	0.93
Pro E	0.3	0.2	0.5	0.9	0.87	0.9	0.84	E

DMM(*I*,*j*)
P_DSM(*I*, *J*)
Comprehensive influence

Figure 7. Organizational evaluation and organization-to-project DSM

5. Conclusion

A systematic methodology in construction integration management has been presented in this paper. The proposed method can evaluate the interaction capability of organizational interface and dependencies influence of project interface. Another significant contribution of this paper is the organization's contribution (percentage of project completion by each organization) to build the organization-to- project DSM.It is significant to combine the interactions among organizational interfaces with the projects interfaces to manage multiple construction projects.The investor of the projectcan establish an effective connection between multiple projects and multiple organizations by using this method. Several aspects of the model presented in this paper merit further examination. First, the organizational interface capability evaluation criteria need to be explained in detail and quantitative. Second, the information iterations in multiple projects will also feedback organizational interactions, which should be further investigated.

6. References

[1] Yang, X. Zhang, and T. Yao, "An overlapping-based process model for managing schedule and cost risk in product development," *Concurrent Eng.: Res. Appl.*, vol. 20, no. 1, pp. 3-16, 2012.

[2] Weshah, N., Ghandour, W. E., Jergeas, G., &Falls, L. C. (2013). Factor analysis of the interface management (IM) problems for construction projects in Alberta. *Canadian Journal of Civil Engineering*, *40*(9), 848-860.

[3] Locatelli, G., Mancini, M., Gastaldo, G., and Mazza, F. (2013). "Improvingprojects performance with lean construction: State of the art, applicability and impacts." *Organize. Technol. Manage. Constr.*: Int. J., 5(2013),775–783.

[4] Shokri, S., Ahn, S., Lee, S., Haas, C. T., & Haas, R. C. G. (2015). Current status of interface management in construction: Drivers and effects of systematic interface management. *Journal of Construction Engineering and Management*, *142*(2), 04015070.

[5] S. D. Eppinger. and T. R. Browning, Design structure matrix methods and applications. *Cambridge, MA: MIT Press, 2012.*

[6] Q. Yang and J. LV, "Project optimization and simulation based on DSM rework risk evaluation matrix," *Systems Engineering-Theory & Practice* (*in Chin.*), *30 (9):*1665- 1671, 2010.

[7] Yang, Q., Kherbachi, S., Hong, Y.S., & Shan, C.(2015). Identifying and managing coordination complexity in global product development project. *International Journal of Project Management* ,*33*(7),1464-1475.

IOP Conf. Series: Materials Science and Engineering **690** (2019) 012016 doi:10.1088/1757-899X/690/1/012016

Reshaping Dormitory by Modular Steel Structure

Yu Luo[1], Dongwei Cheng[2] and Hang Yu[2,1,*]

[1] Tongji Architectural Design (Group) Co., LTD, 1230 Siping Road, Yangpu District, Shanghai, China
[2] College of Architecture and Urban Planning, Tongji University, 1239 Siping Road, Yangpu District, Shanghai, China

Email: yuqizhai1@gmail.com

Abstract: Modular steel structure with a 100% assembly rate, is one of significant methods to realize building industrialization. It has an extensive application prospect due to the short construction cycle and small impact on the surrounding environment.. Based on the characteristics of modular architecture, this paper takes the work of the 2018 Chinese College Students Steel Structure Innovation Competition as an example to discuss the design possibilities of student dormitories. However, due to production mode and transportation conditions, modular building also has some limitations for application. Through research on design, this paper makes the breakthroughs in the design limitations of modular building.

1. Introduction

Different from the traditional construction method, modular building is prefabricated in factory, and only the assembly work is done on site, which greatly shortens the construction period. In addition, it can be reused to improve efficiency and save energy.[1] But the structure system and connection mode of modular building is different from traditional buildings. Therefore, it is of great significance for scholars to study all aspects of modular building.

J. Y. R. Liew, R. M. Lawson et al. studied the modular construction of high-rise buildings, tested the connection modes and relevant reliability, and applied them to engineering practice.[2-3] In recent years the study of modular buildings rise gradually in china. Z. H. Chen, F. Wang, P. F. Zhang et al. conduted experiments to analyze the mechanical properties of the structure system of modular building, also studied the key technologies of envelope system and on-site installation, and the results were applied to related engineering projects such as the dormitory in jinghai district, Tianjin City.[4-5]W. J. Zhu, Z. H. Ren et al. explored the space organization and aesthetic design of modular building, and propose effective strategy for the modular building to break the monotony of space and appearance.[6-7]

The student team of Tongji University participated in the 2018 National College Student Steel Structure Innovation Competition sponsored by China Steel Construction Society.The work "Sifang House" built a multi-storey dormitory with modular steel structure for 1500college students, and won the special prize. By this modular dormitory design, we tryto tap the potential of modular system to facilitate various requirements regarded as favorable architectural functions and space.

2. Competition task

The China Steel Construction Society regularly holds the National College Student Steel Structure Innovation Competition to stimulate students' enthusiasm for structural innovation. Each year, the current advanced structural system and innovative node structure can be seen from the competition works. The competition requirement in 2018 is to build a university dormitory for 1,500 students with prefabricated steel structure. Moreover, some basic conditions are given, such as site type, earthquake resistance type, ground roughness, basic wind pressure, basic snow pressure, structural design conditions, and so on, without specific site restriction. According to the functional requirements of students' dormitory, the buildings should adapt to most regions in the country with the structural system innovation.

3. Design of competition works

This design uses modules that are self-stacking to form four square yards with retreat places that are connected by a common functional ring to meet functional needs and competition requirements. (figure 1).The name "Sifang House" of this work also has multiple meanings, as follows.

1. The university students are from all over the country, this is a dormitory where the students from all over the world. 2. The four square yards in the design meet the dormitory requirement, so this is a four-square dormitory. 3. Because the module building can be quickly disassembled and used in another place, thus this is a mobile dormitory.

Therefore, this dormitory is the harbor for students from all over the world and also can be moved where they are needed. Simultaneously, it can increase or decrease with the number of students increasing or decreasing. Namely, when the number of students decreases, some modules can be removed and transported to another place, but when the number of students increases, some modules would be increased.(figure 2).

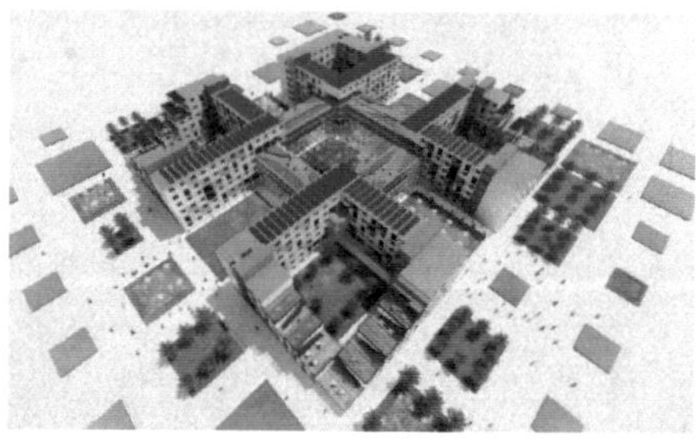

Figure 1. Overall aerial view of the dormitory.

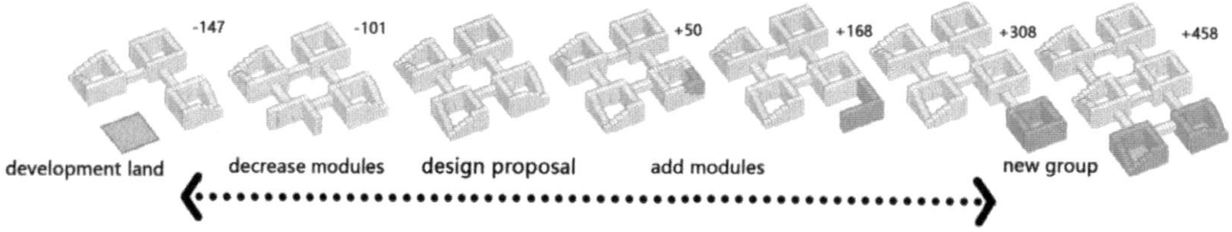

Figure 2. Modular dormitory increases or decreases with the number of students.

The process of design is based on the beginning of a traditional dormitory. 1. In order to improve the living quality, the traditional inner corridor dormitory is transfered into a square yard composed of four outer corridors dormitories. 2. If the square yard is the north-south layout, it is difficult to avoid the problem of sunning the east and west on the left and right sides. Then, the yard is rotated by 45° so that each side of the yard has southern sunshine. 3.Construct a retreat platform at the corner of the yard, which can not only enrich the architectural space, but also most intuitively reflect the features of the module architecture. 4.withcalculation, four courtyards are a suitable layout to meet the living requirements of 1,500 students. 5.A public function ring is introduced to connect the four residential courtyards as a whole, thus creating more public activity space for students.(figure 3)From the process of design generation, it can be seen that the name of the work is not only romantic but also the result of rational analysis based on functional requirements.

4. Design strategies

4.1 Multi-level public space

The introduction of the public function ring is a highlight in the design, furthermore, the public space of the whole program is multi-level. The most important public space at the first level is the function ring in the middle, in which some study rooms and activity rooms can be set up to provide indoor activities for students. The second level of public space is the five courtyards enclosed by the building, which is a foyer for students to enter the indoor space and good outdoor activity space for students. The third level is our retreat platform. These retreat platforms are not empty but also formed by the splicing of framed modules, which can provide students semi-outdoor activity space. The fourth level is to extract several modules to form semi-outdoor activity space between floors.(figure 4).

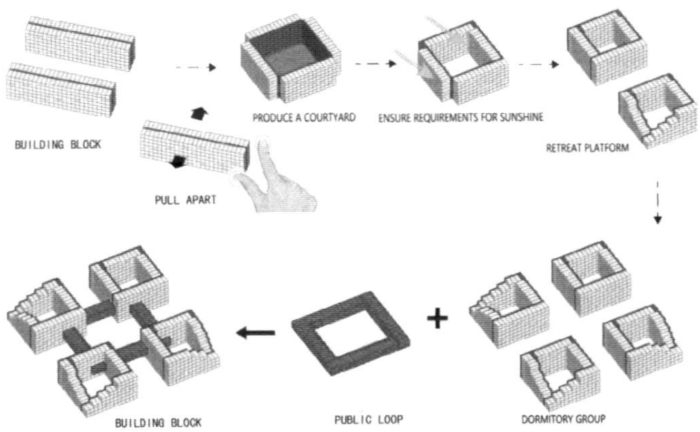

Figure 3. Generation diagram.

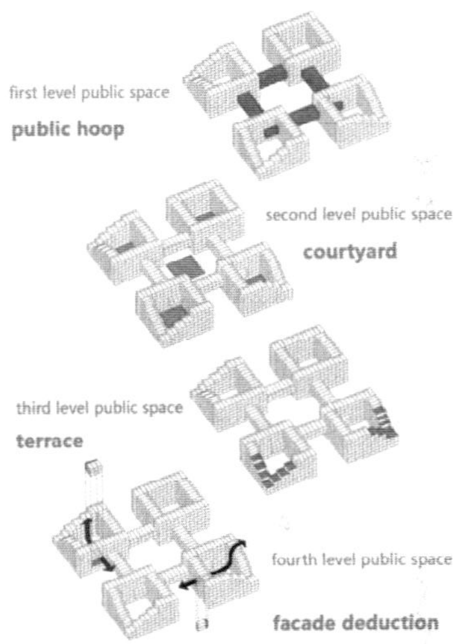

Figure 4. Multi-level public space.

4.2 Module type control

Based on the classification of building functions, the modules used in the entire work can be divided into six types as shown in table 1. Except for the size of the4 and 5 modules are slightly longer, all other modules' size are the same, which is convenient for production, transportation and installation.(figure 5).

Table 1. Module types

	Module type	Module number	Function of the module
1	Barrier-free module	45	Barrier-free dormitory with barrier-free toilet and balcony
2	Standard module	680	Basic dormitory module with a toilet and the balcony
3	Terrace module	22	Terrace for rest and better view
4	Service module	96	Activity room for public activities such as coffee bar
5	Steel framed module	48	Only have structure frame, playing a structural support role
6	Transportation module	116	Containing stairs and elevators

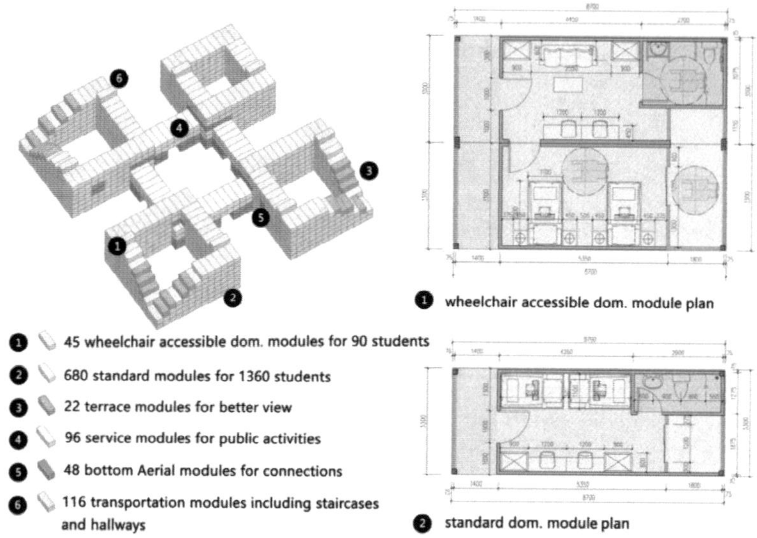

Figure 5. Modular design

4.3 Structure system features

In the architectural construction, this work adopts the light steel keel structure system. With the advantages of light weight, convenient construction as well as rapid construction and installation. The specific construction measures focus on equipment pipeline integration and module connection, and the equipment pipelines are laid in the toilet and walkway to facilitate production and maintenance of equipment. The joint parts between modules should be installed by sealant or patchwork plate on site so as to realize the sealing and the integrity of the building.

The innovation of structural system is also the focus of this competition. Based on the architectural design, the structure adopts the modules' self-stacking way. Simultaneously, the main innovations are the positioning anti-shear double function key and energy dissipation technology. The positioning anti-shear double function key designed in the connection provides locating function during installation to improve the speed and accuracy as well as transfers shear force between modules in service to ensure structural safety. The corrugated steel plate shear wall in the system is smaller than the general steel bracing, not affecting the use of indoor space, moreover, it has the function of energy dissipation and shock absorption. (figure 6)In addition, the structural design not only considers the mechanical properties of the building, but also fully calculates the lifting conditions of each module unit. [8]

The modular building is greatly suitable for the function and scale requirements of the dormitory, in addition to the design of the public space and the innovation of the structural nodes, the design has achieved good results. However, there are still some restrictions which are caused by the characteristics of modules worth to think how to break through.

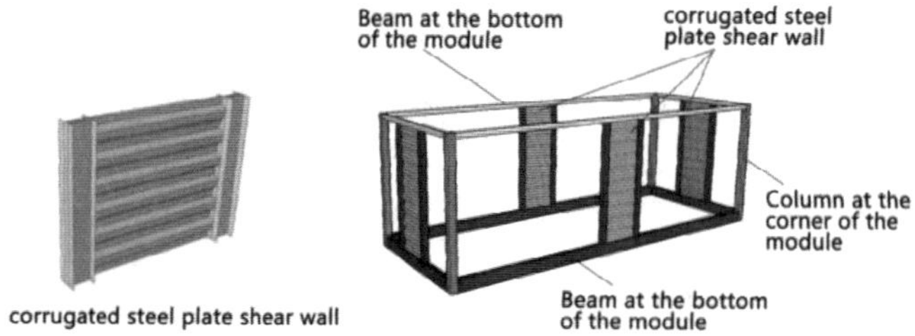

Figure 6. The corrugated steel plate shear wall

IOP Conf. Series: Materials Science and Engineering **690** (2019) 012016 doi:10.1088/1757-899X/690/1/012016

5. Limitations of module building

Although modular buildings have many advantages, currently the costs of construction in most areas are higher than those of traditional construction methods. The combination of modules forms a whole, which could lead to double walls in the whole building interior. Each module structure is independent, the lower module also bear the upper load, the structural steel content would be more than the traditional column and beam system building. There are also some conflicts between the wall structure of the steel structure and the traditional living habits, such as not being able to place the brackets on the wall at will. The most important limitation of modular building is that the size of a single module is limited, which results in difficulty in forming a large space in the building.

Because the production and assembly of module construction have been completed in the factory, a simple unit is transported to the destination, so it is difficult to avoid the size limitation for the transportation. In terms of conventional land transportation, the size standards for cargo transportation are as follows.

The total height is 4.2 meters, the total length is 18 meters, and the total width is 2.5 meters. After deducting the size of the truck, the cargo size is basically 3 meters high, 15 meters long and 2.5 meters wide. Each unit of the modular building is a independent "box" in structure. The columns and beams in the unit not only support its structural integrity but also take into account the other loads in the building as a whole as well as their integrity during transportation and hoisting. Therefore, it is difficult to eliminate some beams and columns of the single unit. The size limitation of each "box" makes it difficult to directly use the modular building to realize the requirements of a large internal space such as a large span and a high height.

6. Limitations break through of modular architecture through design

It is hard to avoid some limitations of modular building, and some can be broken through by design. The height problem of module can be solved, that is, the length of the module can be regarded as a height, which can effectively increase the building height. However, the opening and depth of the building are only about 3 meters, and the module units with such a installation mode should meet the structural stability during transportation and after installation, the structural cost would be further increased.

When the two layers of modules are superimposed, the top plate of the next layer can be removed, and the bottom plate of the upper layer can be appropriately raised, so that the height of the lower building can be appropriately increased. However, the height increase in this way is limited, and the height of the upper module could also be reduced, which makes the modular unit difficult to meet the requirements of large range of height increase. It is still a feasible method to increase the height in a small range to improve the interest of space.

It can be combined with an assembled steel structure system. For example, a hotel building can be constructed with a prefabricated steel structure system on the first floor to meet the needs of the building opening and floor height. The second floor and above use the modular unit to build the room. It is also possible to build a full-height space with fabricated steel structure between two rows of module units to obtain a relatively large space.(figure 7) Such a combination of structural form can well solve the special space requirements, and it is also an ideal way to break through the size limit.

It is possible to meet high space requirements by means of modular unit splicing. The first layer of the public function ring of this competition work has a large entrance space of 6.6 meters wide, 6.6 meters high and 9 meters deep through the stacking of units. The horizontal superposition of the 13.2-meter long modules above the two superimposed modules on the first and second floors is equivalent to the extraction of the following four superimposed modules so as to increase the space size. (figure 8)There should be many similar methods to enlarge the space size and break through the limitation of module unit.

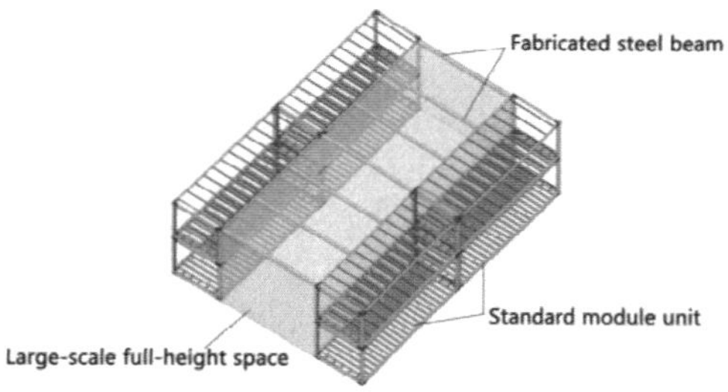

Figure 7. Module combine with fabricated steel structure to build a full-height space .

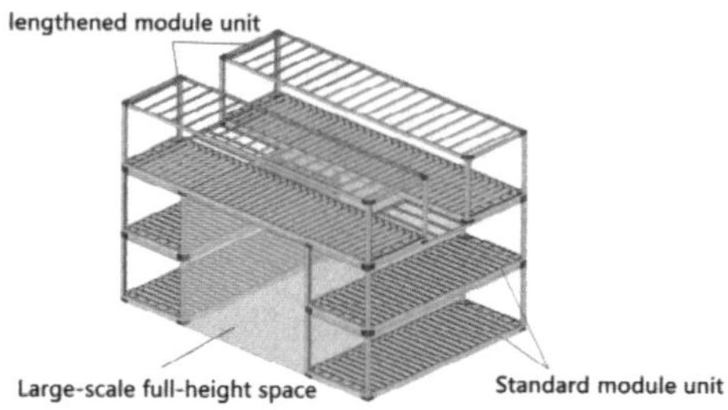

Figure 8. Modular Variable Length Combination to build a full-height space.

7. Summary and rethinking

This article discusses the potential to establish feasibility to accommodate the diversified requirements functionally, socially and aesthetically via modular dormitory design. Though the limited module size places restrictions on the pursue for large space to some extent, delicate engagement and combination of restricted module types and thorough consideration of transportation and installation conditions are thought to contribute to relatively satisfactory feasibility.

8. References

[1] Shou-heng M, Xiao-yu G and Yu-qing Z 2018 *China. J. Architecture & Culture.*115-116
[2] Lawson RM, Ogden RG and Bergin R 2012 *J. Journal of Architectural Engineering.* **18** 148-154
[3] J.Y.R. Liew,Y.S. Chua and Z. Dai 2019 Steel concrete composite systems for modular construction of high-rise buildings*J. Structures*
[4] Pengfei Z, Xizhi Z, Zhihua C, Jiadi L, Xiaodun W and Zechao Z 2015 *Proc. of the 15th National Symp. on Modern Structural Engineering (Kaifeng)* (Tianjin: Tianjin Steel Structure Society)pp 831-836
[5] Fang W, Rongkang Y and Zhifeng L 2017 *Proc. of the 17th National Symp. on Modern Structural Engineering (Tianjin)* (Tianjin: Tianjin Steel Structure Society)pp 764-772
[6] Wenjian Z 2003 *China. J. Architect* 46-52
[7] Zuhua R and Moumeng C 2018 *China. J. Architectural Journal* 10-13
[8] Zhihua C, Xu Z, Yujie Y, YangL and Jiancong H 2018 *China. J. Arch. Technol.* **49**372-76

Research on Body and Architecture

Linxue Li[1], **Qi Zhang**[1] **and Meiting He**[1]

[1] College of Architecture and Urban Planning, Tongji University, Shanghai, 200092, China

*Corresponding author's e-mail: academy@atelierlplus.com

Abstract. The topic of body is closely related to architecture. Based on the Anthony Vidler's and others' study on the relationship between body and architecture, this paper studies the relationship between body and architecture from a historical perspective, including the concept of body in philosophical context, the projection of body in classical architecture, and the metaphor of body in postmodern architecture. The theory of body has had an important influence on the architectural form at the beginning of the development of architecture, and then expanded the understanding of architecture at different stages. Recently, with the environmental issues and sustainable development getting more and more attention, the relationship between body theory and architecture will also face new development potentials and opportunities. Reconsidering architecture from the perspective of body perception and paying attention to the interaction between psychology and environment is an opportunity for the development of body and architecture, which may lead to theoretical and practical guidance of more contemporary significance.

1. Introduction

The topic of body was first mentioned from the field of philosophy and has always been a topic of concern in western culture. The western philosophy starts from paying attention to the human body and takes the body as the starting point to think and understand the objective world.[1] With the further research, the attention and research on the body have also affected the development of architecture. The formation of western architecture, especially classical architecture, was influenced by social culture, politics, economy and philosophy at that time, and it was the result of many factors. As an important concept in philosophy, people's understanding of the body must also be reflected in the buildings and cities at that time. The relationship between the body and buildings is what this article discusses.

Since Vitruvius's *Ten Books on Architecture*, the topic of body, as a representative of the humanistic tradition of architecture, has continuously promoted the development of architecture. Anthony Vidler summarized the development of the relationship between the body and building in his article *The Building In Pain: The Body And Architecture In Post-modern Culture*. The author mentioned that the discussion of architecture and body can be divided into three important stages from the historical dimension: the building as body; the building epitomizing bodily states or, more importantly, states of mind based on bodily sensation; the environment as a whole endowed with bodily, or at least organic, characteristics.[2] Vidler discussed the return of the body in the post-modern architecture.

2. Origin: The body in the philosophical context

The concept of body was first mentioned in philosophy. Plato first mentioned the concept of body in his works. He opposed the body to the soul, thinking that the body is the sinful body. Plato constructed a binary opposition framework for the body and soul. The body was considered as the flesh full of desires which dominated by the soul. Until the Middle Ages, the body was still condemned both by philosophy and religion.

Until the Renaissance, due to the development of humanism, the praise for the beauty of body began to appear. Since the 17th century, philosophy and science defeated the theology, and the body has gradually stepped out of the shackles of theology. However, the main goal of philosophy at this moment is to destroy theology, not to liberate the body. So thhe body is no longer a sinful body, but as an object opposite to the subject, it is dominated by the subject. Descartes' viewpoint of "I think, therefore I am" separates consciousness from body. The body was completely ignored in the subject's thinking and lasted until the 20th century.[3]

In the 20th century, three great scholars freed the body from the state of being ignored and gave it the leading position in philosophy. The first is Merleau-Ponty's phenomenology of body. The second is Durkheim, Moss and Bourdieu's anthropological tradition of social practical body. And the third is Nietzsche's and Foucault's view of historical political body. These three viewpoints are summarized by scholars as individual body (phenomenology of body), social body (the material body is used as a starting point for thinking, and is used as a symbolic language resource to transfer to a discussion of social and moral order) and political body (the control of body by power).

The concept of body always changes with the development of the times. In classical philosophy, due to ancient Greece's praise of the individual body, the focus on the body is more on the individual. In contemporary philosophy, phenomenological research does not go beyond the emphasis on individuals. Merleau-Ponty's phenomenology of body thinks that the body is materialistic, but it is not a simple organism. It is subject-conscious which embodied a concept of "body-subject". In the framework of Durkheim, Moss and Douglas, the body is divided into physiological body and social body, and the body becomes a symbolic system. The same is true of the political body, which is a power system. Since the 19th century, social science has made a great deal of discussions on the body from a social perspective. Marx emphasized the active body, collective and productive body. The study of the body gradually goes beyond the individual level.

The "body" shows more differences in architecture. In classical buildings, the body represents the order of the universe, is a microscopic world, and is the representation of Christ's perfect body. With the development of psychology and empathy theory, the physical and mental research replaces the material level, and the body and experience perception are related. In the study of post-structuralist philosophers, the body represents desire and the force carrying various factors, which are intertwined with space, society and power. The research on the body has promoted architects to think about the essence of architecture.

3. Historical development of the body and architecture

3.1. Body and architecture in classical architecture

3.1.1. Vitruvian Man. Marcus Vitruvius Pollio is recognized as the first person to establish the connection between body and architecture. He proposed the famous "Vitruvius" in *Ten Books on Architecture*. In a square and a circle, one's arm stretches, and the navel is at the center, as a guide to perfect proportions. (Fig.1) Vitruvius believes that the body is a natural form and suggests that this form should be embodied in architecture. He believes that the structure of artistic works should imitate the structural characteristics of the human body.[4]

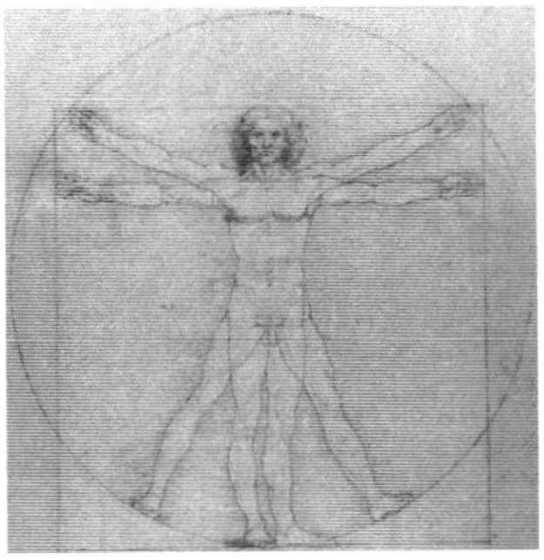

Figure 1. Leonardo da Vinci, Vitruvian Man

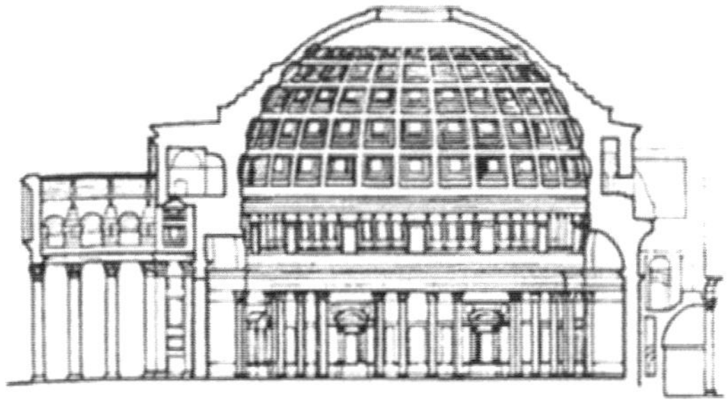

Figure 2. The section of the Pantheon

Vitruvius proposed that the proportion and symmetry of the human body should be applied to buildings to create perfect visual images according to this rule (such as the Pantheon). (Fig.2) This view of analogy between human body and architecture, known as Anthropomorphism, has always existed in the topic of western classical architecture and developed further in the Renaissance. [5]

3.1.2. Filaret. Filaret is the earliest representative figure of anthropometry. He believes that architecture originates from human beings, and therefore evolved from human body, human limbs, and human body proportion. At the same time, he also believes that architecture is not only deduced from the proportion of human body, it is a living organism, and it is a more obscure way to imitate human body.

Filaret further proposed that the body contains cavities, entrances and deep spaces, which form its proper functions. Similarly, buildings also have holes, namely doors and windows. This is true of buildings and cities. In addition, just like the body, buildings and cities can also get sick. The building can get sick and die while it can also be cured by a good doctor. His organic theory implies some ideas of functionalism.

Anthropomorphism was not the main idea of architectural design after Vitruvius, but Filaret extended it to the concept of architectural design. He believed that the form, order and decorate in architecture were inseparable from human appearance, human survival value, human activities and moral values. Filaret creatively interpreted the architectural form with people's moral and ethical class, which is closely related to humanism in the Renaissance.

3.1.3. Francesco di Giorgio Martini. Francesco was a famous architect during the Renaissance, who expressed Anthropomorphism in practice. His concern on body is that he uses the body as a comparison of the city and its components, as well as the form and proportion of the building and its components, from pillars and stigmas to the flat of the church. He proposed that cities have the human's quality, size and shape. In his paintings, a figure is superimposed on the plan of the cathedral and the city, with the navel in the main square. In order to plan a city, he suggested that people should stretch their bodies on the floor, fix a line on their navel, and then draw a circle, which followed the Vitruvius. [6] This provides an analogy not only in proportion but also in shape.

3.1.4. Summary. In Renaissance theories, the physical form of the body is directly projected onto buildings, which not only represent the body, but also express the perfection of the body ideal. Buildings take an organic and autonomous form from the body. Buildings can correct themselves by analogy of their own parts. During the Renaissance, a unique and attractive analogy between body and architecture was formed, which named Anthropomorphism. It's the perfect embodiment of this organic form.

3.2. Body and architecture in postmodern architecture

In the post-modern period, through Nietzsche, Foucault and Deleuze's interpretation of the body, the classical body tradition disappeared, and the body was broken, becoming a fragment mingled in society and space. Many architects saw the vitality of body in this fragment. They want to feed the body back into society and space to create a place where events can happen. It’ s a space which does not stick to the rules and actively challenge the society.[7]

Deleuze believes that the body is the relationship between force and force. If there are two different forces, the body is formed, whether it is social, political, chemical or biological. The body is the difference of force itself. Force is always in the process of generation and flow. [8] For example, the "follies" designed by Bernard Tschumi in the park of La Villette are consistent with Deleuze's concept of body. They do not have a definite shape. They are constantly changing with the help of body movements and are in the process of change. (Fig.3)

The body is shown as movement in Tschumi's buildings, which is the carrier of events in space and acts as a generator and catalyst. At the same time, the body is also the creator of construction violence. Tschumi believes that there are two possibilities for violence in buildings, one is physical violence against a given space, and the other is physical violence in space. Influenced by phenomenology, Tschumi regards the body as the initiator of events and behaviors, but events and behaviors are the main body in Tschumi's space. The violent conflict between space and body is also inevitable. An uncomfortable space or mutating space always brings the body into violence, which breaks the taboo in architecture and makes the body happy. This is the pleasure of violence pursued by Tschumi.

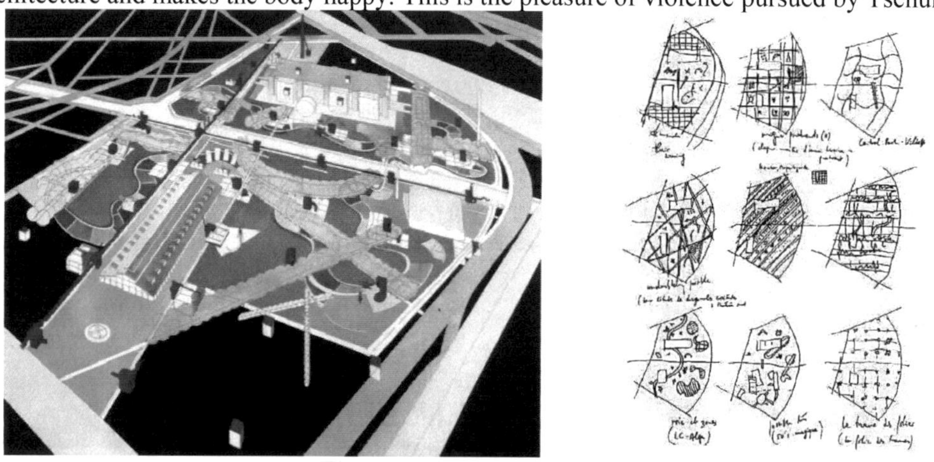

Figure 3. The Park of La Villette

The posture of body in the post-modern period is no longer the perfect body in the classical period. Postmodern characteristics are manifested in the body, such as heterogeneity, instantaneity, split, uncertainty, etc. The post-modern body expands the intersection between architecture and multi-disciplines, also discusses the characteristics of architecture in the post-modern period.

4. Thinking about body and architecture in today's context

Body theory in architecture is closely related to research in humanities and natural sciences. With the increasingly environmental problems, people pay more and more attention to the relationship between man and nature. In the physical research which in the field of natural science, psychology pays attention to the related problems between psychology and environment and the interaction between psychology and environment. As an open system, human beings receive stimulation from the environment to produce perception, and through information processing to produce understanding of the environment or produce behaviors acting on the environment.

The rapid development of science is constantly giving birth to modern people's new understanding of the body. The built environment is seen as an external condition and external expression of human health. This new understanding of the requirements of the building is multi-dimensional. On the one hand, architects hope to change the past built environment by providing the building with plenty of sunlight, flowing air, space for sports, and a hygienic environment. On the other hand, this expression is symbolic, such as erecting the underside of the building to get out of the wet ground. [9]

In today's architecture, the environment is already an unavoidable issue. Rethinking architecture from the perspective of body perception and paying attention to the interaction between psychology and environment may become a direction of thinking about body and architecture. Space affects body perception, and body behavior in turn affects the interaction of space. This is an opportunity for the development of body and architecture, which may lead to theoretical and practical guidance of more contemporary significance. [10]

5. Conclusion

The article discusses the development of the relationship between architecture and body from classical architecture to post-modern architecture. The attention of the post-modern architecture to the body is a re-recognition of Vitruvian's body concept, a denial of the re-negation of the humanist body in the classical period, and a reflection on power and institutions. The theory of body has had an important influence on the architectural form at the beginning of the development of architecture, and then expanded the understanding of architecture at different stages.

After the rapid construction of the city, the relationship between buildings cities and people is alienated and indifferent. The torch of humanism spirit is gradually rekindled. Returning to body is the trend of architectural development. According our research before, we can know that the two organic systems, body and nature, are hostile to modernism, and the relationship between body and architecture is ignored by functional architects. In the 20th century, phenomenology paid attention to the physical experience in life, which leaded the architecture in the 21st century began to pay attention to the sensory experience of the body in space. It started emphasizing the connection between the experience and the real world.

In today's architecture, the environment is already an unavoidable issue. The relationship between body and architecture will also face new development potentials and opportunities. Reconsidering architecture from the perspective of body perception and paying attention to the interaction between psychology and environment maybe can provide more theoretical and practical guidance.

6. References

[1] Yunyi W 2019 The Care of "Heart"—The Study of Architectural Space from the Perspective of "Medical Body" *Beauty & Times* **03** 21-22
[2] Anthony V 1999 The Building In Pain: The Body And Architecture In Post-modern Culture *AA Files* **19** 3-10
[3] Shaojie S, Jiang L 2009 Glimpse of Modern and Contemporary Architecture from Viewpoint of "Body" Architectural Journal 01 27-29
[4] Chaochao C 2010 The Projection of Body in the Classical Architecture *Huazhong Architecture* **28(1)** 145-148
[5] Bin H 2019 Perception of Space *Architectural Journal* **03** 116-122
[6] Shaoyan C, Wei T 2019 Bringing Bodies into Urban Studies: A Critical Perspective for Solving Urban Problems *Urban Planning International* **33(6)** 13-20
[7] Chaochao C 2016 Research on Architecture Ideas of Doller + Scofidio *Huazhong Architecture* **34(12)** 1-4
[8] Jian Z, Xiaoxiao W 2017 Body and Space Teaching Notes *Architectural Journal* **07** 113-119
[9] Pingping D 2017 From Medical Body to a Structure-Related Concept of Environment *Architectural Journal* **07** 14-19
[10] Linxue L 2015 Discourse and Paradigm: Historical Scenario and Contemporary Frontier of Energy and Thermodynamic Architecture *Time+Architecture* **02** 10-16

IOP Conf. Series: Materials Science and Engineering **690** (2019) 012018 doi:10.1088/1757-899X/690/1/012018

Death and Rebirth: Regional Renewal and Development of Industrial Architecture Heritages in Shanghai

Yuda Lu[1,2], Yan Zhao[3] and Yue Shi[4]

[1] College of Architecture and Urban Planning, Tongji University, 1239 Siping Road, Yangpu District, Shanghai, China
[2] College of Architecture, Inner Mongolia University of Technology, 49 Aimin Street, Xincheng District, Hohhot, Inner Mongolia, China
[3] College of architecture, Shanxi University, 92 WuCheng Road, Xiaodian District, Taiyuan, Shanxi, China
[4] Planning and Design Department, China Eco-City Academy Co.,Ltd, Room 1201, Building 7, Interwest Business, No.9 Shoutisouth Road, Haidian District, Beijing, China

Email: 171783720@qq.com

Abstract. Lots of industrial architecture heritages which remain the utility and aesthetic values at the present times are preserved in Shanghai which is one of the earliest industrial cities in modern China. Shanghai as the most exoteric and cutting-edge city among other countries around the world, has the distinct regional feature of avant-garde. Based on the methodologies of field investigations and case studies, the utility and aesthetic value of industrial architectural heritages in Shanghai are integrated with avant-garde for the purpose of reshaping their physical and psychic space.

1. Introduction

Since the 1950s, due to the post-industrial era in European and American, old industrial enterprises were being phased out or moved to the suburbs to decline. In the 1970s, western scholars began to identify these remains as heritage sites. According to the connotation of industrial heritage in "Nizhny Tagil Charter", it can be summarized as an industrial heritage which contains certain historical culture, scientific technology, architectural value, certain social significance and scientific research value. As a part of industrial heritage, industrial architecture heritage pays more attention to the architectural part which contains both material and non-material elements. In terms of material, it includes the remains of production buildings such as workshops, warehouse buildings, ports and mines, as well as buildings of transportation, trade and management equipped with industrial production. It also includes the new technologies and materials in engineering, as well as social and living buildings related to industrial activities. Non-material heritage includes the production process, technology and other forms of cultural expression, as well as cultural heritage in individual memories and behavioral habits.[1]

2. Shanghai and its regional characteristics in modern times

Since the 1840s, with the establishment of foreign concessions and enterprises in Shanghai, the center of southeast in China gradually shifted from Suzhou and Hangzhou to Shanghai. Shanghai became the center of economic, financial, trade and shipping in China.

The development of the old and new Shanghai originated from openness. The modernization of material civilization had brought about the modernization of consciousness. The public had taken the lead in accepting diverse ideological, cultural and artistic influences from foreign countries. In the first half of the 20th century, old Shanghai was forced to modernize. In the second half of the 20th century, new Shanghai took the initiative to open up under the guidance of reform and opening up. In Shanghai, the western advanced technology, thought and theory collided with the traditional Chinese culture of 5,000 years, forming a unique regional culture and spirit with avant-garde and fashion.

Avant-garde means ahead of time, referring to the technology and ideas[2]. The technology and ideas so called "avant-garde" are experiments that attempt to innovate as a whole. It is unorthodox and regarded as deviating from tradition. The avant-garde of technology is the experiment of the latest scientific research. The avant-garde of thought is embodied in the thought mode and the experiment of avant-garde art. Shanghai's advantageous geographical location, developed economy and open attitude make avant-garde develop rapidly and extensively in all walks of life. In the field of architecture, the avant-garde technology refers to the experiment of new techniques, while the avant-garde ideas refers to the experiment of new theories and aesthetics. The development of avant-garde art and ideology has continuously influenced the practice of Shanghai architecture, making it develop rapidly and widely.

According to statistics from Shanghai third-party exhibition information agency, more than 160 exhibitions related to avant-garde art were held in Shanghai in the first half of 2019, while there were also a large number of high-end brand press conferences and creative exhibitions. It can be seen that avant-garde Shanghai has brought the functional space demands of avant-garde exhibition and fashion conference, as well as the prosperity of relevant businesses.

3. Shanghai industrial architecture heritage

3.1. Composition of industrial architecture heritage in Shanghai

For its renewal, it is the necessary step and preparation to clarify the composition of Shanghai's industrial architecture heritage in two aspects of industry type and time.

3.1.1. Industry type composition. Shanghai, as the core of modern industry in China, has a relatively complete industrial type. So far, the number of industrial heritages which are counted, excavated and sorted has reached 300 items, including heavy, military, shipbuilding and transportation industry, etc. which basically cover all kinds of modern Chinese industries. Among them, the textile, paper, printing, food and tobacco industry are the pillar industries of Shanghai. For example, 15 of the 38 existing industrial buildings involving the textile industry are located in Yangpu District, accounting for 40% of the total. This basically reflects the composition of industrial types in this period. Meanwhile, as one of the earliest ports in China, warehousing is also a major industry of Shanghai.[3]

3.1.2. Time dimension. From the first world war to the 1930s, it is a golden period for the development of industry dominated by machinery manufacturing and light industry which includes food, textile, paper, printing and other industries in Shanghai. Based on the investigation of the distribution of the construction time of modern excellent industrial buildings in Shanghai, at least 25 of the 43 excellent modern industrial buildings were built in this period, accounting for about 60% of the total[4].

Based on the analyses, we can roughly understand the type and time composition of Shanghai's industrial architecture heritage. The industry type is relatively varied, mainly in light industry. Its time dimension is main from the first world war to the 1930s. The analyses provide background resources for the renewal and growth of Shanghai's industrial architecture heritage, which are conducive to further study of the renewal.

3.2. Characteristics and values of Shanghai's industrial architecture heritage

Different from the general public buildings, the industrial buildings form distinct symbols in space, structure and appearance, due to the particularity of technological process. In addition to the unique cultural, economic, social and political structure of modern Shanghai, the industrial buildings also have distinct regional characteristics.

3.2.1. Spatial and structural features. The spatial characteristics of industrial buildings are usually determined by industrial process and structure technology. According to different technological requirements, the space and structure types of industrial buildings in modern Shanghai can be divided into general type, large space type and special type. General type: for the industrial building without special requirement in technological process, its space is not novel. It is usually single-story or multi-story, and the structure is no different from general public building, mostly the brick and wood mixed structure and frame structure. Large space type: due to the demands of production, such as storage, power plant and so on, and the progress of structure, factories with large spatial span and strong spatial integrity were built in modern Shanghai, and their main structure is truss. Special type: some industries with special process requirements, such as slaughterhouses, power plants, water plants and docks, the particularities of their technological process are often reflected in the diversity and uniqueness of space, structure and appearances, such as Shanghai Matches House, its gas tank is a large spiral steel structures of four layers, each layer height is 9.3 meters, the top cover is bell jar type, spatial characteristic and symbol are extremely obvious.

3.2.2. Appearance characteristics. Due to the special political and social background of modern Shanghai, a variety of styles coexist in architecture appearance. The appearance characteristics of industrial buildings in modern Shanghai can be roughly classified into several types: traditional, western retro and modern style. Traditional style usually appeared in the workshop without special technological requirements, mainly with the sloping roof and the brick wall. Influenced by western colonization, most of the industrial plants were invested by foreigners in modern Shanghai, so retro western or a combination of Chinese and western style was always used. It was characterized by complex decorative elements, such as moldings, window grilles, arches and orders. The modern style with a simple appearance focused on practical functions and the impact of technological processes on the building.

The unique spatial, structural and external features and the political, social and cultural background of modern Shanghai industrial architecture heritage make it has great value in many aspects. Firstly, Shanghai's industrial architecture heritage has a unique social value. From the cultural perspective, its reuse is the reservation and continuation of a special historical background, including aesthetic, spiritual, cultural, historical, symbolic, authenticity values and evaluation standards. Secondly, its renewal and growth are the transfer and continuation of its economic, material and functional values. In terms of economic value, it is mainly reflected in the yields on heritage projects, which can be divided into three parts: use value, non-use value and externality. What's more, its spatial, structural and external characteristics are in line with Shanghai's urban avant-garde spirit and space requirements for avant-garde exhibitions. The internal features of industrial architecture heritage have incomparable values on exerting and constructing avant-garde in contemporary Shanghai.

4. The avant-garde renewal and growth of Shanghai's industrial architecture heritage

Renewal is the basis of avant-garde growth. The macro-level renewal involves the reconstruction of the relationship between industrial architecture heritage and city, such as function, spatial scale, etc. Micro level renewal refers to the reorganization of its functions, space, structure, appearance and other aspects. In contemporary Shanghai, the renewal of industrial architecture heritage needs to implant the outstanding regional feature of avant-garde into its material and spiritual basis at different levels.

4.1. The macro level

On the foundation of avant-garde, each area in Shanghai owns the different characteristics of avant-garde function and spirit. In the functional reconstruction of industrial architecture heritage at the urban or regional level, it is necessary to accurately locate its functional types and spatial structures according to its regional functional and spatial features, so as to integrate the implanted functional types and structures to improve the quality of urban areas. For example, 1933 Old Mill, is located in the commercial area of Sichuan Road which has a complex urban space structure. According to the original situation, the avant-garde space with multi-scale is implanted, such as fashion performance, display, trading, design, service, business, leisure and so on. As a characteristic product and an important base of living, traveling and consuming for tourists in this region, this kind of implantation interacts with the North Bund of Hongkou District to drive regional economic development.

4.2. The micro level

From the perspective of architecture, the renewal of industrial architecture heritage at the micro level includes the following two aspects: function and space, structure and appearance. Specific renewal strategies need to be reorganized above this two aspects. The design of intervention and non-intervention should be carried out to achieve avant-garde growth.

4.2.1. Function and space. Unique types form unique functions and spaces. The renewal of Shanghai's industrial architecture heritage requires preliminary investigation and classification of its various function and space samples, which are divided into general, special and giant type. Based on the full understanding of the limitations and variability of existing functions and spaces, the feature should be extracted. And then, the original functions and spaces should be replaced adaptively, including translation of old functions and implantation of new functions. In addition, the methods of space integration, separation and expansion can be used to make the update fit with avant-garde spirit. For example, the renovation of 80,000-ton Silo of Minsheng Port, as shown in figure 1, is regarded as giant type. Its original storage space is no longer suitable for advanced grain storage and is abandoned. Its unique high-rise enclosed space can not be adapted to most small scale functions, such as office, catering and so on, but fits with a large scale exhibition function. Therefore, the renewal fully respects the original space characteristics and replaces it with the function of urban pioneer art exhibition, so as to realize the adaptive reuse of function and space.

Figure 1. The interior space of 80,000-ton Silo.

Figure 2. 1933 Old Mill.

4.2.2. Structure and appearance. The unique and varied structure and appearance of Shanghai's industrial buildings make most of them the landmarks of the city. It can be considered that the unique structure and appearance have avant-garde characteristics, so in the process of renewal, they should be sorted out first. Through extraction and analysis, the special structure and appearance should be protected, enhanced and highlighted in the design process, and the original avant-garde should be integrated with the current popular avant-garde, so as to achieve the renewal of sculptural charm. For example, in the renovation of the 1933 Old Mill, the structure was repaired by the integrated pouring of concrete, so that its original and unique ramps, pillars and exterior decoration became the highlights, and became a typical example of making full use of the original structures and exterior features to achieve renewal and growth, as shown in figure 2. For another example, the tube of the 80,000-ton Silo and the big chimney of the Shanghai Museum of Contemporary Art are all reserved, as shown in figure 3 and figure 4. Through the method, this kind of avant-garde reintegration has great significance in the renewal of the structure and appearance of the industrial architectural heritages.

Figure 3. The interior space 80,000-ton Silo.

Figure 4. The big chimney of Shanghai MCA.

5. Case studies of Shanghai industrial architecture heritage renewal

5.1. 1933 Old Mill

1933 Old Mill is a typical example of transforming an abandoned slaughterhouse into an avant-garde fashion center.The main building, built in 1933, was once the largest and most advanced slaughtering factory in the Far East. But with the passage of time, its original function can not match the modern city center function. Building 1, according to the process of slaughtering, formed 26 interleaved corridor bridges in the internal space, as shown in figure 5. Through using the reinforced concrete umbrella-shaped floor without beam, as shown in figure 6, both the space and structure have strong visual impress, and a good match with avant-garde material and spiritual space[5].

Figure 5. The interleaved corridor bridges in the internal space of 1933 Old Mill.

Figure 6. The reinforced concrete umbrella-shaped floor without beam .

The whole transformation is a process of positioning, design and redesign. According to the local function, the space of avant-garde fashion performance, display, design, service, business, leisure and other function with different scales is implanted into Old Mill. At micro level, based on retaining the typical space, structure, spirit and temperament, architect repaired and polished the outdoor and indoor space with cement. With authenticity and discernability, the architect made it form a sculptural unified gray and heavy texture to awaken the body and soul that had been sleeping for decades and to provide a space and platform with avant-garde and fashion temperament [6].

5.2. The Renovation of 80,000-ton Silo of Minsheng Port

Undoubtedly, the most prominent buildings of Minsheng Port are 40,000 tons and 80,000 tons silos, as shown in figure 7. From the perspective of typology, their unique tube sequence structure and space are the result of the combination of historical functions and artistic characteristics. This prominent feature provides citizens a public cultural space with the concept of art, daily life and events. Architect and avant-garde artist presented the three-dimensional traffic between the first and third floors like a ladder which floats outside the silo, as shown in figure 8. With modern materials such as mirror stainless steel and glass, the silo and the ladder of clouds complement each other, creating a work of art that reflects Shanghai's unique avant-garde temperament [7]. As the architect said, "leave more possibilities to the future", the renewal and growth of the old building are highlighted here. This renovation and renewal inspired by artistic and cultural activities has become an effective attempt to renew industrial architecture heritage under typology [8].

Figure 7. Minsheng Port.

Figure 8. The ladder floating outside the silo.

6. Conclusion

As the frontier of China, Shanghai is undergoing a series of avant-garde innovation practices in the transformation of industrial architecture heritage. Industrial architecture heritage with strong historical memory has been redefined. The avant-garde art, creativity and business forms are applied to stimulate the vitality of old buildings, and this kind of theory and practise are being widely discussed and tested. Though the summary and review of the avant-garde growth practice of Shanghai's industrial architecture heritage have relative limitations on the regional and sample dimensions, they provide typology contribution and preliminary attempt to the renewal of industrial heritage.

7. References

[1] Catherine M, Ian S and Jennifer G 2019 *J. Journal of Community Archaeology & Heritage* **6(3)** 173–88
[2] Tianzhong S 1995 *J. Art Critic of Shanghai* **1995(3)** 28
[3] Yupei W 2009 The reuse of Shanghai's industrial heritage: a reference to western experience *D.* (Shanghai: Fudan University) 11–15
[4] Qi H 2007 The protection and reuse of modern industrial buildings in Shanghai *D.* (Shanghai: Tongji University) 70–78
[5] Chongxin Z 2008 *J. Architectural Creation* **2008(03)** 50–57
[6] Chongxin Z 2008 *J. Architectural Journal* **2008(12)** 70–75
[7] Yichun L, Xiaoyi C, Longhai W and Weishi W 2019 *J. World Architecture* **2019(01)** 140
[8] Hong M 2018 *J.Time+Architecture* **201801** 50–57

IOP Conf. Series: Materials Science and Engineering **690** (2019) 012019 doi:10.1088/1757-899X/690/1/012019

Evaluation of the Relationship between the Transit Ridership and Accessibility Variables: A Case Study of the Bangkok Metro Stations

Khin Thiri Kyaw Nyunt[1] and Natachai Wongchavalidkul[2]

[1]Department of Civil and Environmental Engineering, Faculty of Engineering, Mahidol University, Nakhon Pathom, Thailand
[2]The Cluster of Logistics and Rail Engineering, Faculty of Engineering, Mahidol University, Nakhon Pathom, Thailand

Email: khinthirikn@gmail.com

Abstract. Many studies and planning projects related to Thailand metro system is now highly concentrating on the improvements of Transit-Oriented Development (TOD) and the feeder system to ensure smooth transferring for users traveling from their origins to destinations. Leaving off policies related measures, improvements that could immediately focus in order to enhance the smooth transfer services are infrastructure and land-use developments. This study evaluates relationships of accessibility variables, focused on service design and land-usages, and demand of metro ridership. The objective is to improve the understanding of what types of expanding infrastructures or land deployments that could help to support the metro system and encouraging more transit ridership. The existing Bangkok metro stations are the case study area for this research. Gathered data are transport services, roadway infrastructures, and land use. These data are available in the Geographic Information System format and ready to extract as testing variables at each station. Relationship between the number of transit ridership and variables form transport services, network connectivities, and land use are explore based on the correlation analysis. Results from the study found that the number of bus lines, bus stops, and railway stations are associated with the transit ridership, while there is no relationship between ridership demands and ferry services. Further for land-use perspective, the commercial, industrial, and mixed-used area have a significant influence on the ridership demand while there is no signal from the Residential area.

1. Introduction

The accessibility is one of the significant indicators in transport planning. It helps planners define the connectivity status between land use and transportation in their study areas. Planners can use accessibility indicators to define the effectiveness of changes in the system infrastructures or policies. Hence, improving transport accessibilities will directly advance both service quality and the number of users [1][2].

The existing metro system in Thailand is now operating with the only total of 100 kilometers of services, approximately. However, within the year of 2035, the service will be more than 464 kilometers, covering both Bangkok and its vicinity areas. The developments of this future system are well-progressing in terms of the central service infrastructures or the metro itself, except its surrounding system. Hence, many studies and planning projects related to Thailand metro system highly concentrate on the improvements of Transit-Oriented Development (TOD) and the feeder system to ensure smooth transferring for users traveling from their origins to destinations. Leaving off policies related measures, improvements that could immediately focus in order to enhance the smooth transfer services are infrastructure and land-use developments. This study evaluates relationships of accessibility variables, focused on service design and land-use, and demand of metro ridership. The objective is to improve the understanding of which infrastructure or land deployment that could help to support the metro system and encouraging more transit ridership.

2. The measure of accessibility in transportation

There are four types of accessibility measures classified in the literature, including infrastructure-based, location-based, person-based, and utility-based measures [3][4]. The infrastructure-based accessibility measure focuses on time, level of congestion, and average travel speed in the transport network. The location-based accessibility measures the spatial distribution of opportunities or activity demands in the area, i.e., the number of jobs within 30 mins travel time. The person-based measures focus on the activities in which an individual can participate within a given time. Finally, the utility-based accessibility measures the individual-level benefits from land use or transport improvements.

Additionally, four fundamental components of accessibility are transport, land-use, temporal, and individual components. The transport components are transport infrastructure, passenger demands, and service determinations of the infrastructure, including times, costs, and efforts to use such services. The land-use component reflects the land use system consists of the amount, quality, and spatial distribution activities found at each destination (jobs, shop, and home). The temporal component relates to the availability of activities at different times. The individual component is the peoples' needs, abilities, and opportunities in using such the services. Table 1 presents the relationship between accessibility measures and accessibility components for classifying the type of accessibilities [5][6][7].

Table 1. Accessibility measures and components.

Measures/ Components	Transport component	Land use component	Temporal component	Individual component
Infrastructure based measures	Average travel time; Travel speed; Level of congestion		Peak hour period	Trip based stratification (e.g., Homework, business trips)
Location-based measures	Travel time or costs between locations of activities	Spatial distribution of opportunities in space (e.g., Job per zone)	Travel time and costs may vary in time (e.g., Between peak hours and off-peak hours)	Stratification the population (e.g., by income, education level)
Person-based measures	Travel time	Distribution of opportunities in space	Time constraints for activities and time available for activities participation	Accessibility is analysed at an individual level
Utility-based measures	Travel time or costs between locations of activities	Distribution of opportunities in space	Travel time and costs may vary in time (e.g., between peak hours and off-peak hours)	The utility is derived for population groups level or at the individual level

Source: [7]

As prior discussions, this study focuses on the improvements of accessibilities by paying attention to infrastructure improvements and land use developments. Hence, as present in table 1, the evaluations in this study will focus on the incorporation of the location-based measure and the land-use component. There is a relation between land use and accessibility. If accessibilities change, land use will also relatively change. For example, improving transit coverage services will also enhance business development and residential land use in those development areas. For location-based, due to its availability and simple, the most used measure found in researches is distance (i.e., distance from home to public transport services). There are four types of location-based accessibility measures found in the literature. These measures are (1) direct distance measurement, (2) travel time, distances, or cost contour, (3) potential accessibility measures (i.e., concentrations of jobs and population), and (4) balancing factor measures [6][7].

3. Case study area

The case study area is the existing metro stations in Bangkok, Thailand. Five operating lines with 79 stations are currently opening for services. This study considered only the urban area stations, which has a total of 64 stations. Figure 1 presents the distribution of the selected study stations.

Gathered data are transport services, roadway infrastructures, and land use. These data are available in the Geographic Information System format and ready to extract as analyzed variables at each station. The number of ridership data (Y) is the most existing data from the service operators. For the independent variables (Xi), besides its availabilities, types of variables are from recommendations and usages in the literature related to transit-oriented development. All data are in the Geographic Information System (GIS) format and ready for extract as the testing variables. Additionally, collected data concentrating for those spatially existed around 800 meters around the studied stations. Section 5 discussed these variables again in more details.

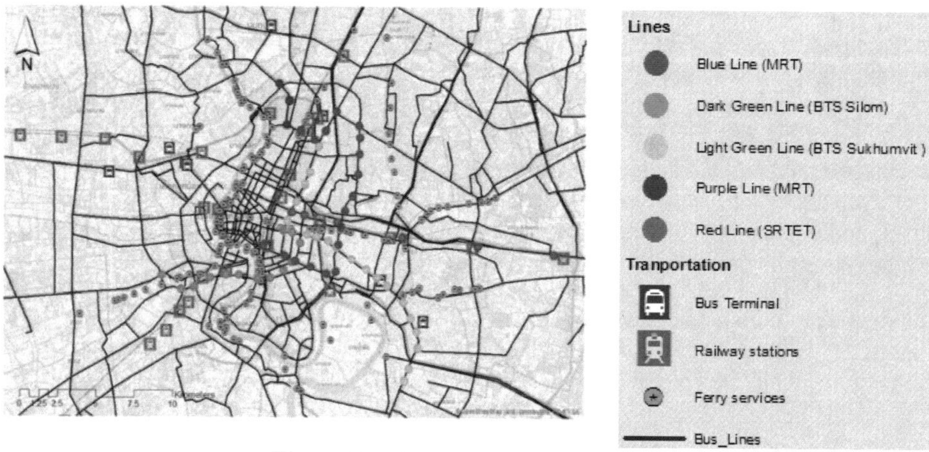

Figure 1. Bangkok metro stations area

4. Methodology

Pearson's correlation analysis is used to analyze the relationship between variables and select the critical variable in order to explore the significant accessibility measures for enhancing future transit service and development [8]. The rule of thumb for analyzing variable correlation in this study are as follows:

1. The statistical significant is when $0.01 < P < 0.05$
2. If $P < 0.01$, it means that p values associate between two variables is strongly significant
3. If $P > 0.05$, p-value associated between the two variables is not significant
4. R-value is normally greater than 0.6 will be considered as a strong relationship between variables

5. Data collections

The collected data for this study has two categories (transportation connectivity and land-use). The transportation connectivity incorporates of eight transportation variables includes: the distance to municipal public service facilities, the distance to landmark and retail services, the distance to public transport stations, the number of ferry services, the number of bus lines, the number of bus stops, the number of railway stations, and the street network. The first three variables represent the Euclidean distance of the rail system to the major facilities or landmarks. Additionally, two main variables for land-use are the development scale of residential and other uses of commercial industrial and mixed-use development. The variables represent development scale in terms of floor area (square meters) development. Table 2 presents the list of collected data in this study.

IOP Conf. Series: Materials Science and Engineering **690** (2019) 012019 doi:10.1088/1757-899X/690/1/012019

Table 2. List of collected Data.

Groups	Data	Variables	Units
Transportation Connectivity	Distance to Municipal Public Service facilities	X_{11}	Meters
	Access to landmark and retail services	X_{12}	Meters
	Public transport accessibility	X_{13}	Meters
	Number of ferry services	X_{14}	Ferries
	Number of bus lines	X_{15}	Lines
	Number of bus stops	X_{16}	Stops
	Number of railway stations	X_{17}	Stations
	Street network	X_{18}	Meters
Land use	Residential area	X_{21}	Sq-m
	Commercial, Industrial and mixed-used area	X_{22}	Sq-m

6. Results and discussions

From Pearson's correlation analysis, figure 2 presents the correlation matrix for the variables considered this study. From figure 2, many variables have a high correlation to others. The significant correlation between variables are:

- Access to landmark and retail services (X_{12}) has a high correlation with Distance to municipal public service facilities (X_{11})
- Public transport accessibility (X_{13}) has a high correlation with Commercial, Industrial and mixed-used area (X_{22})
- Public transport accessibility (X_{13}) has a high correlation with Street network (X_{18})

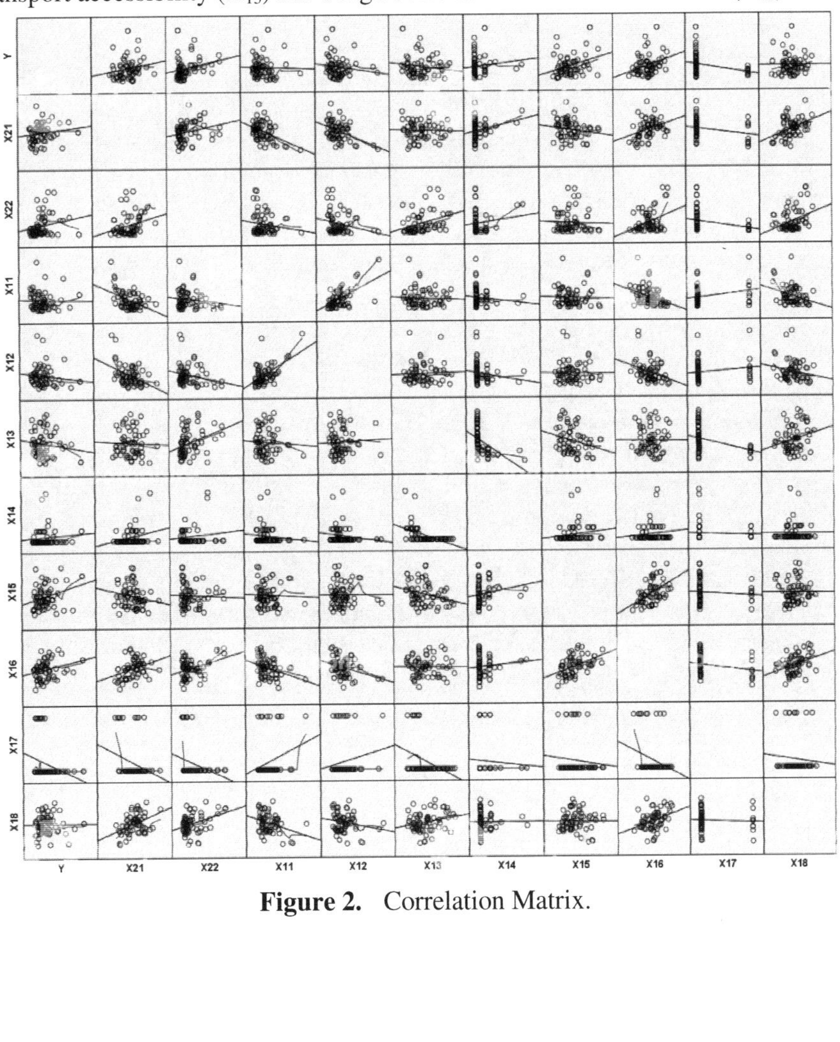

Figure 2. Correlation Matrix.

This preliminary correlation results, it presents that accessibility currently exists to support the commercial, industrial, and mixed land use. Besides, Table 3 shows the results of the correlation between transit ridership (Y) and accessibility variables (X_{ij}). The results found that when considered R-value, only a few variables have R-value greater than 0.2, and no variables have R-value higher than 0.4. However, some variables show significant results of the p-value, which meant that these variables are a part of influencing to the ridership volumes. These variables are Number of bus lines (X_{15}), number of bus stops (X_{16}), number of railway stations (X_{17}), and Commercial, industrial and mixed-used area (X_{22}).

From a simple correlation analysis between selected accessibility measures and transit ridership in this study, the results are as follows. The number of bus lines (X_{15}) and the number of bus stops (X_{16}) within the 800-buffer area are positively associated with the transit ridership, while the relationship between ridership demands and the number of railway stations (X_{17}) is negative. On the other hand, the results cannot find the relationship between ridership demands and ferry services (X_{14}).

Further for land-use perspective, the commercial, industrial, and mixed-used area (X_{22}) have a significant influence on the ridership demand while there is no signal from the Residential area (X_{21}). In this case, it is an unexpected result since most land development in Bangkok has a high residential concentration around the transit stations. The reason must be that more people in Bangkok still choose to drive for their work trips, while the metro is a choice for other trip types.

Table 3. Pearson's correlation analysis of associability variables and ridership.

Variables	Y	X_{21}	X_{22}	X_{11}	X_{12}	X_{13}	X_{14}	X_{15}	X_{16}	X_{17}	X_{18}
Correlation coefficient	1	.160	.217*	-.026	-.132	-.148	.105	.310**	.285*	-.271*	.003
P-value		.104	.043	.419	.149	.122	.204	.006	.011	.015	.490

7. Conclusions

For future research, further study is also progressing to comprehend more variables such as population density and some other design variables such as roadway configuration. Since results of correlation matrix found that there is a severe multicollinearity problem in the stock of variables, linear regression may not be fit for the future model in this study. Other modeling technique such as random parameter estimations and the neural network may be suited to the problems.

8. References

[1] Cheng J, Bertolini L and le Clercq F 2007 Measuring sustainable accessibility *Transportation research record* vol 1 pp 16 - 25.

[2] Manout O, Bonnel P and Bouzouina L 2018 Transit accessibility: A new definition of transit connectors *Transportation Research Part A: Policy and Practice* 113 pp 88-100.

[3] Handy S L 2002 Travle behavior-land use interactions: An overview and assesste of the research. IN: in pereptual motion: Travel behavior research opportunities and application challenges.

[4] Geurs K T and Van Wee B 2004. Accessibility evaluation of land-use and transport strategies: review and research directions *Journal of Transport geography* vol 12 no.2 pp 127-140.

[5] Curl A, Nelson J D and Anable J 2011 Does accessibility planning address what matters? A review of current practice and practitioner perspectives *Research in Transportation Business & Management* vol 2 pp 3-11.

[6] Geurs K T and Ritsema van Eck J R 2001 Accessibility measures: review and applications. Evaluation of accessibility impacts of land-use transportation scenarios, and related social and economic impact *RIVM rapport* 408505006.

[7] Breheny M J 1978 The measurement of spatial opportunity in strategic planning *Regional Studies* vol 12 no.4 pp 463-479.

[8] Chuan C L and Penyelidikan J 2006 Sample size estimation using Krejcie and Morgan and Cohen statistical power analysis: A comparison *Jurnal Penyelidikan IPBL* vol 7 no.1 pp.78-86.

IOP Conf. Series: Materials Science and Engineering **690** (2019) 012020 doi:10.1088/1757-899X/690/1/012020

Validation of Bridge Scour HEC RAS Model Using Propagation Error Analysis

Billy[1], Dwinanti R Marthanty[1] and Jessica Sjah[1]

[1] Civil Engineering Department, Faculty of Engineering, Universitas Indonesia, Depok, 16424, Indonesia

Email: Billysujana@gmail.com

Abstract. The scouring phenomenon is one of the causes of bridge damage in the world. This scouring phenomenon has various kinds such as local scour and contraction scour. There are various equations for calculating pier scouring, including Colorado State University equations, Frochelin and others. A computer program is needed to facilitate the calculation of pier scouring, HEC-RAS based on Finite Difference Method was chosen to simulate pier scouring problem. Up until now, there has not been any research to quantify the uncertainty of bridge scour HEC RAS based of experiment. The objective of this study to validate the pier scouring calculation by comparing the expected value of pier scouring results of HEC-RAS with an experiment and CSU equation. The HEC-RAS was evaluated using experimental data that had been carried out by Shukri in 2017. Furthermore purpose of this study is determining which the variable that is sensitive to the depth of scouring. The sensitive variables are determined by performing sensitivity analysis of the variables on HEC-RAS using the Pearson correlation coefficient method and regression analysis. Error propagation is done to get the uncertainty value from pier scouring due to the influence of sensitive variables.

1. Introduction

Scour is the result of grinding due to water flow, which lifts material from the riverbed and riverbanks around the piers and bridge abutments [1]. The scour phenomenon needs to be considered because it will reduce the carrying capacity of the soil due to eroded soil. If the bridge piers or foundation is not strong enough to withstand the force of the water flow and the bridge load, a bridge piers is collapsed or shifted.

There are two types of scour, namely local scour and contraction scour. This study focuses on the phenomenon of local scour which involves grinding the material around the piers, abutment, and river embankment. This results in a decrease in the surface of the riverbed which causes a decrease or shift in the piers or abutment. The scouring pattern that occurred on the pier was horseshoe, with the flow of water that was bigger make deeper into the scour [2,3]. Local scouring around the bridge piers occurs due to the flow of water held by the piers, horizontal flowing water will be held back by the bridge piers and form a rotation system [4,5].

This numerical study quantified the depth of pier scouring by using the HEC-RAS based on experiment conducted by Shukri in 2017 [6]. The equation of scouring is the result of momentum conservation law by applying the conservation law of mass, in 1992, Richardson found an equation for calculating depth pier scouring called CSU Equation [7]. The CSU equation was equation that used in the HEC-RAS.

The purpose of this research is to validate the HEC-RAS to calculate the pier scouring depth. The sensitivity analysis for pier scouring caused by some variables was evaluated. The studied variables were velocity of water flow, flow depth, and discharge. The expected value of pier scouring depth from the simulation results of HEC-RAS will be compared with one obtained by the experimental results [6] and CSU equation.

2. Pier scouring

Scouring was the result of erosive actions of flowing water, by dredging and carrying away material from the bottom and riverbanks and from around the bridge pier and bridge abutments [7]. Different types of material will produce different scour, such as cohesive soil (clay) will be more difficult to erode compared to non-cohesive (sand) soil.

Local scour in the pier is a function of the characteristics of riverbed material, basic configuration, flow characteristics, fluid properties, and pier geometry. The material characteristics are granular or non-granular, cohesive or non-cohesive, erodible or rock that cannot be defined. Granular bed material ranges from small to large stones and it is characterized by *D*50 and rough sizes such as *D*84 or *D*90.

CSU equation is the recommended equation by the US highway administration transportation department to calculate the depth of pier scouring:

$$\frac{y_s}{y_1} = 2.0K_1K_2K_3\left(\frac{a}{y_1}\right)^{0,65} Fr_1^{0.43} \tag{1}$$

where, y_s = scour depth, y_1 = flow depth directly upstream of the pier; Fr_1= Froude number; K_1= correction factor for pier nose shape; K_2 = correction factor for angle of attack of flow; K_3 = correction factor for bed condition ; a = pier width.

Richardson has determined the values of the correction factors needed in the CSU equation [7]. For the pier shape correction factor (K_1), the value ranges is from 1.1 - 0.9, depending on the shape of the pillar. Flow direction correction factor (K_2) is determined based on flow angle, pillar length and pillar diameter, with a range of 1 - 5. The bed material correction factor is determined by the size of the grain, with a value of 1.1 - 1.3.

3. Method

There are 4 stages to complete this study. First stage is the process of collecting data for the pier scouring input process in the HEC-RAS application, which is in the form of river geometry data, water flow data, and soil data. The data is taken from the results of the experiment [6]. There are also hypothetical data for cross sections of rivers and bridge structures.

The second stage is performing numerical simulation by using HEC-RAS and data obtained in stage 1. Numerical modelling is done with 3 scenarios. Scenarios are performed to see variables that have a significant effect on the depth of pier scouring, and it is estimated that there are three studied variables that have the effect on pier scouring depth from several input parameters needed in HEC-RAS modelling. The first scenario is the variation of flow depth with the variation of 0.01 – 0.1 m, the second scenario is water velocity with the variation of 0.045 – 0.54 m/s, and the third scenario is flowrate with the variation of 1.57 – 25.1 dm^3/s. The scenarios are carried out on three different pile diameter variations which are 5.5 cm, 7.5 cm, and 11 cm.

The third stage is carried out by performing sensitivity analysis of the simulation results of three scenarios on HEC-RAS. Sensitivity analysis is performed to see the effect of each variable on pier scouring depth by using the Pearson method about correlation coefficient data processing. Correlation coefficient (R) is a value that indicates the strength or absence of a linear relationship between two variables. There is a correlation relationship created by Pearson, so that is called Pearson correlation. The value of R is -1 to +1, where when the value of R approaches 1 or -1, then the relationship between the two variables is strong, whereas if the value of the variable approaches 0, then the variable has a weak relationship. From the Pearson value, it can be determined that the depth of pier scouring can be called sensitive to the variable varied [8]. This is the equation of Pearson:

$$R = \frac{n\sum_{i=1}^{n} X_i Y_i - \sum_{i=1}^{n} X_i \sum_{i=1}^{n} Y_i}{\sqrt{n\sum_{i=1}^{n} X_i^2 - \left(\sum_{i=1}^{n} X_i\right)^2}\sqrt{n\sum_{i=1}^{n} Y_i^2 - \left(\sum_{i=1}^{n} Y_i\right)^2}} \quad (2)$$

The final stage is done by examining the error analysis to compute the uncertainty from the depth of pier scouring due to the influence of three studied variables. Error analysis is a study and evaluation of uncertainty in measurement. In a measurement, it is obtained the best estimation results of observations and there is also an uncertainty of observations whose results are a range or expected value of measurement. The above statement, can be express by the following equation:

$$x = x_{best} \pm \delta x \quad (3)$$

$$\delta q = \sqrt{\left(\frac{\partial q}{\partial x}\delta x\right)^2 + \ldots\ldots + \left(\frac{\partial q}{\partial z}\delta z\right)^2} \quad (4)$$

The expected value of pier scouring depth will be obtained from the simulation results using HEC-RAS. From the result, it can be compared the simulation results using HEC-RAS with the experimental results and literature entered between the upper limit and lower limit of the result or not. If it enters the interval, the result can be accepted.

4. Result and discussion

Modelling is done by entering the needed data in the HEC-RAS. Pier scouring depth is obtained after running the bridge scour window based on the number of variations in scenario parameters and 3 variations diameter of the bridge pillar. There are 111 pier scouring model in HEC-RAS, that divided to 30 running of scenario 1, 36 model of scenario 2, and 45 model of scenario 3. In scenario 1, with variations in the value of water depth (y_1), and the input value of water velocity (V) 0.225 m/s. In scenario 2, with variations in the value of the water velocity (V), and the input flow depth value (y_1) 0.03 m. In table x shows the results of 15 experiments in scenario 1, with variations in the value of water discharge (Q). The depth of pier scouring (Ds) simulation of HEC-RAS with variations of pile diameter: A = 5.5 cm, B = 11 cm, and C = 7.5 cm. In figure 1 there are graphics of scour depth (Ds) versus the variated parameters.

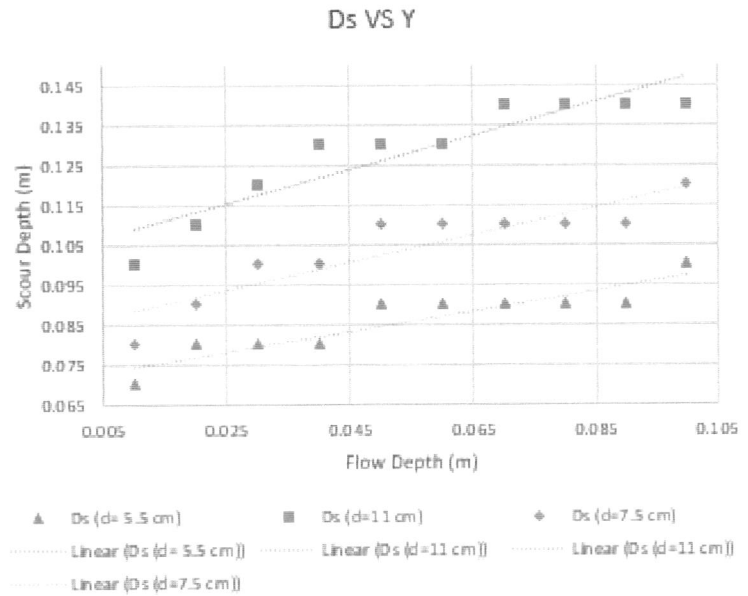

(a)

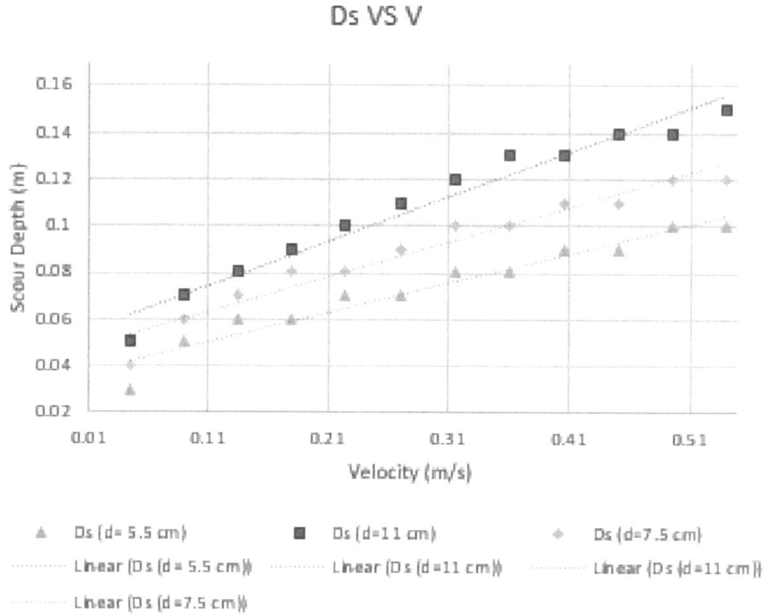

(b)

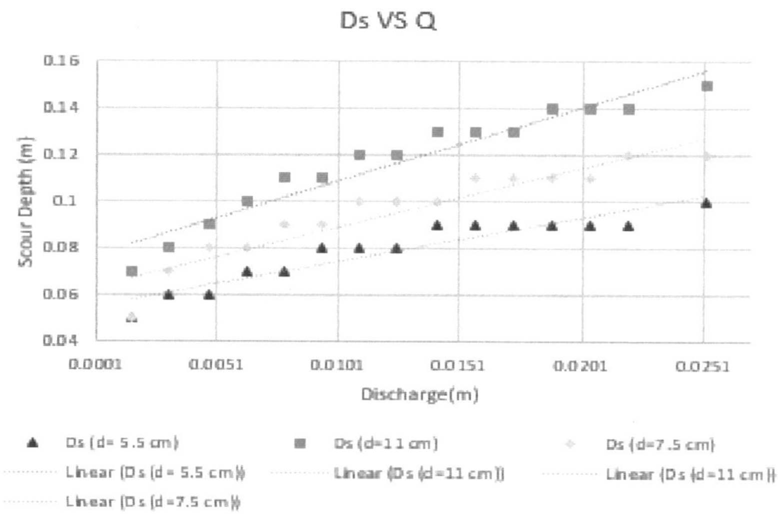

(c)

Figure 1. Scour depth vs: (a) Flow Depth; (b) Velocity of water flow; (c) Discharge from running HEC-RAS

4.1. Sensitivity analysis

Based on the results of data processing from the 3 simulation scenarios performed, it can be concluded that the pier scouring parameter (Ds) tends to be the most sensitive to the water flow velocity variable (V), because and its coefficient correlation (R) approaches 1 from 3 scenario. The coefficient correlation is calculated using equation (2) for 3 variation scenarios. Sorted from the highest to the lowest value, the parameters that have a significant effect on pier scouring depth (Ds) are the velocity of the water flow (V) with $R = 0.97$, the discharge (Q) with $R = 0.94$, and into the flow depth (y_1) with $R = 0.91$.

4.2. Error analysis

The velocity of water flow (V) is the parameter that has the most influence on the depth of pier scouring (Ds), followed by discharge (Q), and flow depth (y_1). The three variables have a coefficient correlation (R) which is close to one and is included in the range 0.8-1, which is categorized as having a strong relationship. Therefore an error propagation is calculated to find the uncertainty value (δDs) from the depth of pier scouring (Ds) due to 3 variables.

The uncertainty value (δDs) can be searched using the slope and standard deviation values of the variables that affect the depth of pier scouring (Ds) from each scenarios. There are 3 uncertainty values (δDs) for 3 variations in pier diameter. The expected value from the depth of pier scouring (δDs) is obtained by adding or subtracting the value of the best estimate with the uncertainty value (Ds), equation (3). Then the uncertainty value (δDs) is obtained and the expected value of pier scouring depth (Ds) for the 3 diameter variations is as follows, the uncertainty (δDs) of type A, B, and C, are 0.024 m, 0.038 m, and 0.031 m.

4.3. Validation

Pier scouring is calculated in 3 different diameters of 5.5 cm, 11 cm, and 7.5 cm. The expected value of pier scouring were obtained in the form of adding the results of running HEC-RAS plus or less with the uncertainty (δDs) that had been calculated use equation 3. The pier scouring depth (Ds) of the experimental results and the CSU equation will be compared with the expected value of pier scouring (Ds) as a result of the HEC-RAS simulator as a validation process. If the value of pier scouring (Ds) by the experimental result and the CSU equation is in the expected value, the results of HEC-RAS are acceptable, if is not then the simulation results of HEC-RAS are not accepted.

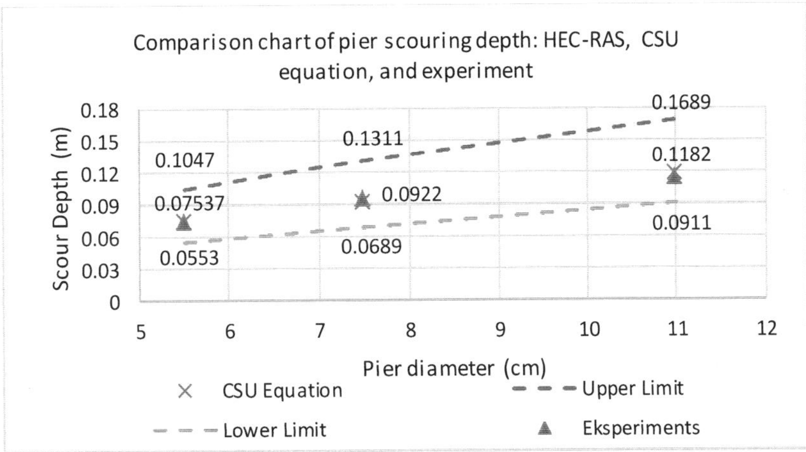

Figure 2. Pier scouring depth comparison chart (Ds): HEC-RAS and CSU equation

Based on figure 2, the depth of pier scouring (Ds) as a result of the CSU equation and the experiment entered into the expected value is the result of the HEC-RAS simulation propagation error. The simulation performed on the HEC-RAS results is acceptable when compared with the results of CSU equation calculations and experiments. The expected value for pier type A, B, and C are 0.08 ± 0.0247 m, 0.13 ± 0.0389 m, and 0.1 ± 0.0311 m. This shows that the simulation results of HEC-RAS are acceptable.

5. Conclusion

The biggest influence on pier scouring depth (Ds) is the velocity of water flow (V), with a value of $R = 0.97$. The second largest influence is the flow discharge (Q) with a value $R = 0.94$. The last is the flow depth (y_1) with $R = 0.91$. The depth value of pier scouring is sensitive to changes in variables, with the most sensitive sequence being the velocity of water flow (V) [m], water flow discharge (Q) [m3 / s], and water depth (y_1) [m] (inflow). These parameters affect the results of scouring depth, so uncertainty is sought, as the tolerance value of the depth scouring results of HEC-RAS. Uncertainty is added to the scour depth value of the HEC-RAS results to produce the expected value.

The HEC-RAS generates scouring depth values as the best estimate value. This value is added to the uncertainty of the results of the error propagation so, it becomes expected value of the depth of scouring. The expected scour depth of HEC-RAS for pile type A, B, and C are 0.08 ± 0.0247 m, 0.13 ± 0.0389 m, and 0.1 ± 0.0311 m. The results of the scouring depth were compared with the results of experiments and CSU equations were compared to expected value. The depth value of pier scouring from the CSU equation and the experiment is entered into the expected value so that the HEC-RAS is can be used to find the pier scouring depth.

6. References

[1] Amerson, L. Z., Lagasse, L., & Clopper, P. (2012). *Evaluatin Scour at Bridges Hydraulic Engineering Circular 18, Fifth Edition.* Washington, D.C: Federal Highway Administration, U.S. Department of Transportation.

[2] Azmi, M. (2014). *Simulasi Gerusan Lokal Pilar Jembatan Menggunakan Model Fisik Sediment Flow Channel HF 302.* Depok: Universitas Indonesia

[3] Guan, D., Chiew, Y. M., Wei, M., & Hsieh, S. C. (2019). Characterization of horseshoe vortex in a developing scour hole at a cylindrical bridge pier. *International Journal of Sediment Research*, 118-124.

[4] Breuser, H., & Raudkivi, A. (1991). *Scouring.* Rotterdam: IAHR Hydraulic Structure Design Manual.

[5] Das, S., Ghosh, S., & Mazumdar, A. (2014). Kinematics of Horseshoe Vortex In a Scour Hole around Two eccentric Triangular Piers. *International Journal of Fluid Mechanics Research*, 296-317

[6] Shukri, M. T. (2017). Experimental Study of Local Scour Depth around Cyclindrical Bridge

[7] Richardson, E. D. (1990). *Highways in the River Environment, FHWA-HI-90-016.* Washingtom, D.C.: Federal Highway Adminstration, U.S. Department of Transportation.

[8] Corder, G., & Foreman, D. (2009). Comparing Variables f Ordinal or Dichotomous Scales: Spearman Rank-Order, Point-Biserial, and Biserial Correlations. *Nonparametric Statistics for Non-Statisticaans; A step-by-step Approach.* Pier. *World Academy of Science, Engineering and Technology International Journal of Civil and Environmental engineering Vol:11*

Acknowledgement

This research is funded and supported by DRPM Universitas Indonesia through Indexed International Publication Grant for Student Final Project (Hibah PITTA 2019). The Number of Contract is: NKB-0743/UN2.R3.1/HKP.05.00/2019.

IOP Conf. Series: Materials Science and Engineering **690** (2019) 012021 doi:10.1088/1757-899X/690/1/012021

A Multistage Fuzzy Comprehensive Evaluation Model based on AHP for Construction Enterprises through Value Chain Analysis

Daoyi Guo[1]

[1] College of Architecture and Urban Planning, Shandong Jianzhu University, Jinan, China

Email: guoguoguo8880@gmail.com

Abstract. The comprehensive performance evaluation for engineering project has received much consideration from academic and corporate over the past decade, developing rapidly and becoming a main area of research for dealing with the complexity of construction management. This paper develops a multistage fuzzy comprehensive evaluation model with systematic and synergy-focused appraisal system from a value chain analysis perspective by using Analytical Hierarchy Process (AHP), to help the industrial practitioners conduct the performance evaluation in a fuzzy environment by handling the vagueness and subjectivity with parameterized linguistic values. The work presented in this paper may help managers and business professionals not only to evaluate the efficiency and effectiveness of the management but also to distinguish the important criteria for creation in value chain, to remain competitive in the market.

1. The performance management of construction enterprises through value chain analysis

Due to the fierce competition in the market the impact of the evolution of global value chains in the internet era and increasingly stringent requirements of engineering quality, the construction industry's profit margin has declined gradually. In Netherlands, Finland and other European regions, the annual growth rate of market demand for construction enterprises is below 5%, which is lower than the annual growth rate of labour costs. This paper uses the value chain analysis method to tease out a series of activities set by construction enterprises aiming at increment of value, extending forward to customers, backward to suppliers and service providers, to deliberate the value creation potential of each link relative to the final product, as shown in Table 1.

It can be seen that the use of information technology and optimization of critical operational processes is the key to realizing enterprise strategy in the Internet Era [1]. By flexibly applying information technology, giving full play to its leverage role and multiplier effect, enterprises can enhance their competitiveness. And through the integration and coordination of the upstream and downstream of the supply chain, the production cycle of construction products can be shortened, so as to reduce the costs of transportation and inventory [2]. The major sources of value creation can be found in the manifold links of the value chain [3].

Table 1. The value process production of construction enterprises

	Enterprise infrastructure	**Human Resource Management**	**Technology Development**	**Purchase**
Internal Activities	Organizational Structure	Recruitment	Method Statement	Procurement Scheme
	Management System	Selection & Training	Construction Schedule	Material Supply Plan
	Corporate Culture	Appointment	Technical Documents	Equipment Debugging
	Budget Accounting	Staff Appraisal	Support & Service	Maintenance Facilities
External Activities	**Marketing and Sales**	**Internal Logistics**	**External Logistics**	**Production Management**
	Market Linkage	Receiving Inspection	Supply Chain	Project Supervision
	Marketing System	Dispatching	Logistics Cost Control	Acceptance Check
	Marketing Research	Store Management	Activity Management	Assurance System
	Reasonable Quotation	Data Filing	Schedule Management	Warranty Service

2. The performance appraisal index system of multistage fuzzy comprehensive evaluation model

The evaluation model of construction projects should be a combination of a series of progressive evaluation activities running through the whole project cycle [4], guided by the unified concept and involving multiple subjects [5]. The evaluation and the project management support each other. Therefore, based on the above value chain analysis, this paper constructs a comprehensive evaluation index system of construction enterprise projects for each link of the project, as shown in Table 2. There is some incomparability due to the different influences of various factors on the engineering project [6], so the index system is divided into two parts: project operation evaluation and project synergy evaluation (including the collaborative communication with customers). Each part has three levels of indicators, including qualitative indicators which are divided into five evaluation levels (as shown in Table 3) and quantitative indicators obtained by the calculation model.

Table 2. The Comprehensive evaluation index system

Primary Indicator	Secondary Indicators	Three-Level Indicator
Evaluation of Project Operation	Project Planning	Project Scope Management
		Project Risk Management
	Project Control	Schedule Control
		Quality Control Coefficient*
		Safety Control Coefficient*
	Project Closure	Project On-Time Completion Rate
		Acceptance Rate
		Delivery of Products
Evaluation of Project Synergy	Collaboration with Customers	Customers Satisfaction Degree
		Timely Delivery Rate*
		Problem Processing Time
	Financial Synergy	Capital Supply Rate *
		Cost Deviation Rate*
		Financial Capital Coordination
	Collaboration with Stakeholders	Criterion Coordination
		Labour Supply Rate *
		Material Supply Rate *
		Equipment Supply Rate *

Table 3. Evaluation level of qualitative indicators

Indicator	Level 1	Level 2	Level 3	Level 4	Level 5
Scope management	Vague	Not very accurate	Basically accurate	Accurate	Accurate and quantifiable
Risk management	Empirical risk aversion	Preliminary strategy	Preliminary management	Come up with alternatives	Time and fund reserve
Schedule control	Schedule delay	Not timely and inefficient	Basically, timely and effective	Efficient control with quantification	Ahead of schedule
Completion rate	A<20%	20%≤α<40%	40%≤α<60%	60%≤α<80%	A≥80%
Acceptance rate	A<20%	20%≤α<40%	40%≤α<60%	60%≤α<80%	A≥80%
Delivery of products	Extensive redevelopment	Partial redevelopment	Basically deliverable	Delivered	Successfully delivered
Satisfaction degree	A<20%	20%≤α<40%	40%≤α<60%	60%≤α<80%	A≥80%
Processing time	Poor (t≥8hours)	Fair (t<8hours)	Average (t≤5hours)	Good (t≤3hours)	Excellent (t≤1hours)
Financial coordination	Poor	Fair	Average	Good	Excellent
Criterion coordination	Poor	Fair	Average	Good	Excellent

In order to eliminate the problem that various influencing factors cannot be compared due to different statistical calibres and dimensions; factors are converted into dimensionless percentages to obtain the calculation model of quantitative indicators reflecting project performance as follows:

(1) Safety Control Coefficient, which indicates whether the project has set up safety protection measures and organized work in accordance with the requirements of safe production. The Quality Control Coefficient reflects the ability of the quality assurance measures of the project under construction to meet the quality requirements of the project. The calculation method of the Quality Control Coefficient of the project is the same as that of the Safety Control Coefficient. The calculation model of the two is:

$$S=\sum_{j=1}^{n}\lambda_{ij}\chi_{ij}=(\lambda_{i1},\lambda_{i2},\ldots,\lambda_{in})(\eta_{i1}/\eta_{0\,i1},\eta_{i2}/\eta_{0\,i2},\cdots,\eta_{in}/\eta_{0\,in})^{T} \quad (1)$$

λ_{ij}--- the contribution rate of each safety/quality measure to the safety/quality of construction projects, which can be comprehensively evaluated through the analysis of historical data and the fuzzy scores of the importance of the construction of the measures; η_{ij} --- The actual assessment of safety/quality measures, the quantitative ones can be scored by calculation , the ones that cannot be quantified are comprehensively scored by the expert matrix method; $\eta_{0\,ij}$ --- the requirements of the safety/quality measures of the project, the method of obtaining is equivalent to η_{ij}.

(2) Capital Supply Rate reflects the difference between the funds available for construction in progress and the actual demand for construction projects, and the distribution of engineering funds among tasks. The calculation formula of Capital Supply Rate is:

$$C=\sum_{j=1}^{n}\eta_{ij}/\sum_{j=1}^{n}\eta_{0\,ij} \quad (2)$$

η_{ij} --- the actual acquisition of funds for each sub-project; $\eta_{0\,ij}$ --- the funds required for each project. The sensitivity of each sub-project to the funds can be additionally calculated by analysing the historical data(λ_{i1}, λ_{i2}, ..., λ_{in}), providing the basis for the next step fund allocation

(3) The Material Supply Rate reflects the difference between the raw materials that can be used in construction projects and the actual construction needs of the project. The calculation formula of Material Supply Rate is:

$$M=\sum_{i=1}^{n}\lambda_{ij}\frac{\eta_{ij}+v_{ij}}{\mu_{ij}} \quad (3)$$

η_{ij} --- the amount of reserves of various raw materials; v_{ij} --- the quantity of expected purchases; μ_{ij} --- the amount of raw materials required for the current project; λ_{ij} --- the impact factor of this raw material on the progress of the project.

(4) Equipment Supply Rate reflects the difference between the mechanical equipment that can be used in construction projects and the actual construction of engineering projects. The Labour Supply Rate reflects the difference between the amount of labour that can be used in construction projects and the actual labour required for construction of the project. The calculation formulas for the two supply rates are the same as follows:

$$E=\sum_{i=1}^{n}\lambda_{ij}\frac{\eta_{ij}{}^{*}v_{ij}}{\eta_{0\,ij}{}^{*}v_{0\,ij}} \tag{4}$$

η_{ij} --- the number of engineering machinery and equipment / the quantity of labour force; $\eta_{0\,ij}$ --- the number of mechanical equipment / the quantity of labour force required for each sub-project construction; v_{ij} --- the efficiency of each mechanical equipment / each unit of labour; $v_{0\,ij}$ --- The optimal efficiency (or quota) of each equipment/ each unit of labour; λ_{ij} --- the impact factor of each mechanical equipment/ each unit of labour on the construction of the project.

(5) Timely Delivery Rate's calculation formula is:

$$D=\frac{\eta}{\eta_0} \tag{5}$$

η--- times of timely delivery; η_0 --- total delivery times.

(6) Cost Deviation Rate is the basis for calculating the economic efficiency of capital construction investment plan, construction organization plan and structural technical plan by calculating the quantity of labour, material and machinery consumption in a certain unit of products according to the requirements of itemized projects and structural components.

$$L=\sum_{i=1}^{n}\lambda_{ij}\frac{\eta_{0\,ij}{}^{*}v_{0\,ij}-\eta_{0\,ij}{}^{*}v_{ij}}{\eta_{0\,ij}{}^{*}v_{0\,ij}} \tag{6}$$

$\eta_{0\,ij}$ --- the amount of completed work for each sub-project; v_{ij} --- the actual cost of each sub-project; $v_{0\,ij}$ --- the budget quota for each sub-project; λ_{ij} --- the proportion of each sub-project cost to the total cost of construction.

3. The multistage fuzzy comprehensive evaluation model

the construction project is a multi-factor complex system with a certain ambiguity, consisting of both quantitative and qualitative indicators in the project performance evaluation index system [7]. Therefore, by using single-factor fuzzy evaluation related to the evaluation objects to constitute the evaluation matrix under the action of the weight vector, the multistage fuzzy comprehensive evaluation model for project performance was established [8]. The general steps are shown as follows:

3.1. Set indicator sets and evaluation sets at all levels

There are two finite universes: $U=\{X_1, X_2, ..., X_n\}$ and $V=\{Y_1, Y_2, Y_3, Y_4, Y_5\}$, where U represents a collection of multiple evaluation indicators of the comprehensive evaluation system of the construction enterprise project, called the indicator set; V is a collection of multiple judgments. It is called an evaluation set, using five levels of excellent, good, medium, average, and poor. According to the comprehensive evaluation index system of construction enterprise projects established in the previous section, the three-layer fuzzy comprehensive evaluation index set is established as follow

$U=\{X_1$(Evaluation of Project Operation), X_2(Evaluation of Project Synergy)}

$X_1=\{X_{11}$(Project Planning), X_{12}(Project Control), X_{13}(Project Closure)}

$X_2=\{X_{21}$(Collaboration with Customers), X_{22}(Financial Synergy), X_{23}(Collaboration with Stakeholders)

$X_{11}=\{X_{111}$(Project Scope Management), X_{112}(Risk Management)}

$X_{12}=\{X_{121}$(Schedule Control), X_{122}(Quality Control Coefficient), X_{123}(Safety Control Coefficient)}

$X_{13}=\{X_{111}$(Project On-Time Completion Rate), X_{112}(Acceptance Rate), X_{123}(Delivery of Products)}

$X_{21}=\{X_{211}$(Customers Satisfaction Degree), X_{212}(Timely Delivery Rate), X_{213}(Problem Processing Time)}

$X_{22}=\{X_{221}$(Capital Supply Rate), X_{222}(Cost Deviation Rate), X_{223}(Financial Capital Coordination)}

X_{23}= {X_{231}(Criterion Coordination), X_{232}(Labour Supply Rate), X_{233}(Material Supply Rate), X_{234}(Equipment Supply Rate)}

3.2. Determine the weight coefficient of the index system with AHP method

Generally, the influence of each indicator in the indicator set on the evaluated object is inconsistent, so the weight distribution of the indicator set is required as a fuzzy vector on the evaluation set V [9]. In theory, the method of determining the weight of indicators is divided into two categories: subjective weighting method and objective weighting method which's disadvantages are the cumbersome calculation and poor practicality. So, the analytic hierarchy process (AHP) method combining qualitative analysis with quantitative analysis with strong operability is selected [10], using less quantitative information to conduct a scientific quantitative analysis of problems that are difficult to directly measure [11].

The analytic hierarchy process refers to grouping the system indicators according to the dominance relationship abiding by scale sequence to form an ordered hierarchical structure [12]. Therefore, the relative importance of the indicators in each level should be determined by means of pairwise comparison judgment. The weight assignment reference table is shown in Table 4:

Table 4. Reference table for weight assignment

Importance degree	P_n
Indicator $X_{n-1}$is as important as the indicator X_n	1
Indicator $X_{n-1}$is slightly more important than indicator X_n	3
Indicator $X_{n-1}$is significantly more important than indicator X_n	5
Indicator X_{n-1} is more important than indicator X_n	7
Indicator X_{n-1} is extremely important than indicator X_n	9
The median of the above judgments	2,4,6,8

The synthesis is then performed within the hierarchical structure to obtain the relative importance of the indicator to the target [13]. The index evaluation matrix can be obtained by asking relevant experts to assign the indicators of each level of the evaluation index system. And the weight distribution of each level of the evaluation index system, $A=(a_1,a_2,\cdots,a_n)$, $A_n \in F(U)$ would be drawn subsequently by analysing the assignments through the data processing system software (DPS) [14]. In the formula, A_n is the weight coefficient corresponding to the nth index X_n and satisfies $A_1+A_2+...+A_n=1$. The weight set is as follows:

$$A_1=(A_{11}, A_{12}, A_{13})$$
$$A_2=(A_{21}, A_{22}, A_{23})$$
$$A_{11}=(A_{111}, A_{112})$$
$$A_{12}=(A_{121}, A_{122}, A_{123})$$
$$A_{13}=(A_{131}, A_{132}, A_{133})$$
$$A_{21}=(A_{121}, A_{212}, A_{213})$$
$$A_{22}=(A_{221}, A_{222}, A_{223})$$
$$A_{23}=(A_{231}, A_{232}, A_{233}, A_{234})$$

3.3. Establish the multistage fuzzy evaluation matrix

3.3.1. The first level comprehensive evaluation. By making a comprehensive evaluation of each indicator in A_i(i=1,2) and look for the membership relationship of each indicator in the index set U to each element in the evaluation set V, the paper established a membership function. The evaluation value of the quantitative indicator in this paper can be obtained according to the calculation result (evaluation score = indicator calculation value/indicator historical optimum value), and the evaluation value of the qualitative index can be judged by experts according to the indicator evaluation table, thus obtaining the membership $Y_{ijtk}(t=1,2,\cdots,m;k=1,2,\cdots,p)$ of the index X_{ijt} belongs to the *k*th comment. Each indicator constitutes a fuzzy evaluation vector $R_{ij}=(r_{ij1}, r_{ij2}, \cdots, r_{ijp})$, the fuzzy evaluation matrixR_{ij} composed of all single-factor fuzzy vectors is as follows:

$$R_{ij}=\begin{vmatrix} r_{ij11} & r_{ij12} & \cdots & r_{ij1p} \\ r_{ij21} & r_{ij22} & \cdots & r_{ij2p} \\ \vdots & \vdots & & \vdots \\ r_{ijm1} & r_{ijm2} & \cdots & r_{ijmp} \end{vmatrix} (i = 1,2;j=1,2,3)$$

Since the comments in the evaluation set are not absolutely positive or negative, the comprehensive evaluation can be regarded as a fuzzy set on V, which is recorded as$B_{ij}=(b_{ij1}, b_{ij2}, \cdots, b_{ijp})\in F(V)$. The fuzzy transform T_R: $F(U)\to F(V)$, $A\to A\times R$ could be Constructed, so that the ternary equation (U, V, R) constitutes a fuzzy comprehensive evaluation mathematical model. Therefore, the weight distribution A_{ij} of the input index layer X_{ij} is:

$$A_{11}=(A_{111}, A_{112})$$
$$A_{12}=(A_{121}, A_{122}, A_{123})$$
$$A_{13}=(A_{131}, A_{132}, A_{133})$$
$$A_{21}=(A_{121}, A_{212}, A_{213})$$
$$A_{22}=(A_{221}, A_{222}, A_{223})$$
$$A_{23}=(A_{231}, A_{232}, A_{233}, A_{234})$$

Therefore, $B_{ij}=A_{ij} * R_{ij}=\left(b_{ij1},b_{ij2}, \cdots, b_{ijp}\right)(i=1,2;j=1,2,3)$

3.3.2. The second level comprehensive evaluation. By taking each B_{ij} as an indicator, the fuzzy comprehensive evaluation matrix of $R_i=(r_{i1}, r_{i2}, \cdots, r_{ip})$ is as follows:

$$R_i=\begin{vmatrix} B_{i1} \\ B_{i2} \\ B_{i3} \end{vmatrix}=\begin{vmatrix} r_{i11} & r_{i12} & \cdots & r_{i1p} \\ r_{i21} & r_{i22} & \cdots & r_{i2p} \\ \vdots & \vdots & & \vdots \\ r_{ij1} & r_{ij2} & \cdots & r_{ijp} \end{vmatrix} (i=1,2)$$

From the previous section, the weight distribution A_i of the index layer X_i to the target layer is obtained.

$$A_1=(A_{11}, A_{12}, A_{13})$$
$$A_2=(A_{21}, A_{22}, A_{23})$$

Secondary evaluation vector $B_i=A_i * R_i=\left(b_{i1},b_{i2}, \cdots, b_{ip}\right)(i=1,2)$ would be obtained.

3.3.3. The third level comprehensive evaluation. By taking each B_i as an indicator, the fuzzy comprehensive evaluation matrix of $R=(r_1, r_2,)$ is as follows:

$$R=\begin{vmatrix} B_1 \\ B_2 \end{vmatrix}=\begin{vmatrix} r_{11} & r_{12} & \cdots & r_{1p} \\ r_{21} & r_{22} & \cdots & r_{2p} \\ \vdots & \vdots & & \vdots \\ r_{i1} & r_{i2} & \cdots & r_{ip} \end{vmatrix}$$

From the previous section, the weight distribution A of the index layer X to the target layer is obtained.

$$A_{11}=(A_{111}, A_{112})$$

The three-level evaluation vector would be obtained: $B=A * R=\left(b_1,b_2, \cdots, b_p\right)$. in accordance with the principle of maximum membership degree, it could be drawn that $B_0=max\left\{b_1,b_2, \cdots, b_p\right\}$, leading to the comprehensive evaluation result for the project is B_0.

4. Conclusion

This paper adopts a multi-level comprehensive fuzzy evaluation method to realize the performance evaluation of engineering projects through the value chain analysis. It should be pointed out that in addition to the evaluation of the project after completion, the model can also be used in the implementation process of the engineering task. By making objective predictions on the construction conditions in the task, the ideal project progress can be calculated depending on circumstances, analysing the impact of factors such as capital supply, equipment supply, raw material supply and other factors on the operation in the original plan [15]. Furthermore, due to the complexity of project management of construction enterprises and the long-term value-added management, the model should focally adapt to the actual situation of construction enterprises to enhance the implement ability and theoretical significance of the model for better application to the construction management in future.

5. References

[1] Keča L J, Keča N and Rekola M 2013 *Value chains of Serbian non-wood forest products* (International Forestry Review vol 15) p 315-335.

[2] Vrijhoef R and Koskela L 2000 *The four roles of supply chain management in construction* (European journal of Purchasing & Supply Management vol 6) p 169-178.

[3] Crain D W and Abraham S 2008 *Using value-chain analysis to discover customers' strategic needs* (Strategy & Leadership vol 36) p 29-39.

[4] Gao L and Hailu A 2012 *'Ranking management strategies with complex outcomes: An AHP-fuzzy evaluation of recreational fishing using an integrated agent-based model of a coral reef ecosystem* (Environ. Model. Softw vol 31) p 3–18.

[5] William H, Xiaowei X and Prasanta D K 2010 *Multi-criteria decision-making approaches for supplier evaluation and selection: A literature review* (European Journal of operational research vol 202) p 16-24.

[6] Roberto A, Iris T, Kenneth W and James H 2003 *Value stream analysis of a re-engineered construction supply chain* (Building Research & Information vol 31) p 161-171.

[7] Caplar R and Kulisic P 1973 *A performance evaluation model by integrating fuzzy AHP and fuzzy TOPSIS methods* (Expert Systems with Applications vol 37) p 7745-7754.

[8] Yücenur GN, Vayvay Ö and Demirel NÇ 2011 *Supplier selection problem in global supply chains by AHP and ANP approaches under fuzzy environment* (The International Journal of Advanced Manufacturing Technology vol 56) p 823–833.

[9] Zadeh L A 1965 *Fuzzy sets* (Information and Control vol 8) p 338–353.

[10] Armillotta A 2008 *Selection of layered manufacturing techniques by an adaptive AHP decision model* (Robotics and Computer-Integrated Manufacturing vol 24) p 450-461.

[11] Chan F T S, Kumar N, Tiwari M K, Lau H C W and Choy K L 2008 *Global supplier selection: A fuzzy-AHP approach* (International Journal of Production Research vol 46) p 3825-3857.

[12] Kahraman C, Cebeci and Ruan D 2004 *Multi-attribute comparison of catering service companies using fuzzy AHP: The case of Turkey* (International Journal of Production Economics vol 87) p171-184.

[13] Rabelo L, Eskandari H, Shaalan T and Helal M 2007 *Value chain analysis using hybrid simulation and AHP* (International Journal of Production Economics vol 105) p 536-547.

[14] Calantone R J, Benedetto C A and Schmidt J B 1999 *Using the analytic hierarchy process in new product screening* (Journal of Product Innovation Management vol 16) p 65-76.

[15] Darmawan MA, Putra MPIF and Wiguna 2014 B *Value chain analysis for green productivity improvement in the natural rubber supply chain: a case study* (Journal of Cleaner Production vol 85) p 201-211。

IOP Conf. Series: Materials Science and Engineering **690** (2019) 012022 doi:10.1088/1757-899X/690/1/012022

The Rules in Reconnecting the City Fabrics: A Morphological Study of the Regeneration Case

Y. Zhai[1]

[1] College of Architecture and Urban Planning, Tongji University, Shanghai, 200092, China

Email: yuqizhai1@gmail.com

Abstract. In the recent 30 years, Shanghai experienced fast urbanization which landscape shows various types of conflicts among the modern forms and the traditional forms. The insertion of new typologies in the traditional district introduces new urban characteristics and dynamics, which are not typically associated with the traditionally framed cities. The constructions of substantial industrial and commercial projects have caused the demolition of traditional urban forms. Urban designers aspire to revitalize the traditional neighborhood by combining urban fabrics without destroying the structure of the historic neighborhood. This paper attempts to exam the effectiveness of the urban network, the continuity of public structures and operational interactions between new projects and the local context through the study of a "failed" regeneration project in Shanghai, China. Consequently, this paper raised rules in reconnecting the city fabrics towards differential urbanism.

1. Introduction

The lack of connectivity, social diversity, or loss of social and economic dynamics have raised the issues of fragmentation, segregation, and marginalization (Pinto and Remesar, 2012)[1], especially in the traditional central area of metropolitan. Since the end of the 20th century, traditional compounds in the central cities of China had been replaced by modern developments. These new projects, supported by the state government, were constructed rapidly with little consideration of the public structure. These typical Chinese-pattern large projects included a central commercial complex, transportation hubs, and gated communities. The insertion of these new simple typologies into the traditional district are not typically associated with the traditional forms, which leads to discontinuity of public space, few interaction between the new project and local context and an inactive spatial network. Currently, the insertion of these new-projects had brought a series of problems in the traditional district.

2. Theoretical approaches

The fragmentation reality of the metropolis is never a new topic. Actually, during the rapid development of the post-war rebuilt of the cities, architects and urban theorists had already set the debate on how to combine the modern large-scale infrastructure and development projects with the more compact and small-scale traditional forms (Khaled A, 2017) [2]. Rowe and Koetter illustrated that even the most ancient cities were built in incremental fragments. In an appendix, they showed that streets work as urban elements in the cities (Rowe and Koetter, 1975)[3]. Rowe's exuberance alienated Sir Leslie Martin, the head of Cambridge, to try to apply mathematical and proportional systems to the city. Also, here started Christopher Alexander. He summarized that the problem of the modern city is in its tree-pattern connection, while the natural city connection is semi-network. In the perspective of pattern and scale, Alexander advocated the "language" of buildings and paths in everyday urbanism.

Therefore, through the morphological "language" of the natural city, everyday social interactions can be combined into an organizational framework (Alexander, 1977)[4].

Subsequently, his former colleague and follower, Salingaros stated that it is possible to combine the modern form with the traditional form through the rebuilt of the city network. Salingaros interpreted that link, node, and hierarchy are the essential elements for the city to contain differential urban patterns (Salingaros, 1998)[5]. In addition, he believed that the smaller the scale, the more effective the pattern connected to the human beings. Therefore, the human mind can combine patterns into group from small to large. So, based on the fractal theory, he then proposed the urban network theory. There were scholars referred similar notion in the concept of "nodal structure" (Gosling and Maitland, 1984) [6] or "armature" (Buchanan, 1988)[7] to reinforce and distinguish the important elements and structures that combine several urban networks (Roberts et al., 1999)[8]. Other major classical streams of this topic include the work of Jane Jacobs (1960), Camillo Sitte (1965), Edmund Bacon (1974) and Zukin. They also emphasized the importance of the divergent and complimentary of the traditional spatial patterns. These approaches also linked to the conception of "re-urbanization", developed by Portas et al. in 2011, where he stated that it is vital to combine activities inside urban fabrics (Pinto A J and Brandão A L, 2015)[9]. Even in the traditional district, plots vary in size and shape, probably as a consequence of a historical and predominantly unplanned process of formation and continuous adaptation (Remali, 2015)[10]. However, private houses usually occupy smaller plots than public buildings in the traditional district which is essential to generate public network (Remali, 2017)[11].

To conclude, the core argument of this paper is as follows:

First, the procedural of the small to large network brings the most active personal connections at the beginning, and build successive patterns based an intuitively accepted base; Second, the public space plays a crucial role in the urban structure and social appropriation. Thus, through the study of the structure of the public structure in these areas, it is possible to understand the specific urban network in it; Third, we understand the public space as collective part of the city, forming a coherent whole and supporting morphological, economy, social and environmental issues of different natures (Carmona et al., 2003[12]; Pinto A J and Brandão A L, 2015[9]). Therefore, public spaces cannot be studied as a sum of isolated elements but as a network (Pinto and Remesar, 2012)[13].

3. The study of the "failed" regeneration project in the 1990s in Hongkou District

The north Sichuan Road District is located in the Hongkou district in Shanghai, China (Figure 1; Figure 2). In the late 1990s, there was a regeneration of the North Sichuan Road district following the national commercial planning. However, when the new construction of the elevated highways, subways, large commercial centres, and large-scale modern gated residential blocks were planted inside the traditional neighborhood, they have not brought a long-term revitalization. (Figure 3; Figure 4).

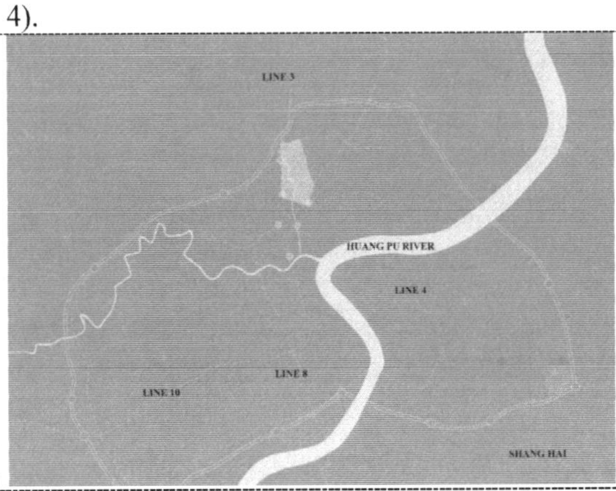

Figure 1. The location of the North Sichuan Road in central Shanghai. Source: Redraw from the studio work.

Figure 2. The photograph of the traditional residential house (left) and the new projects (right) in the Sichuan Road district. Source: photographed by author.

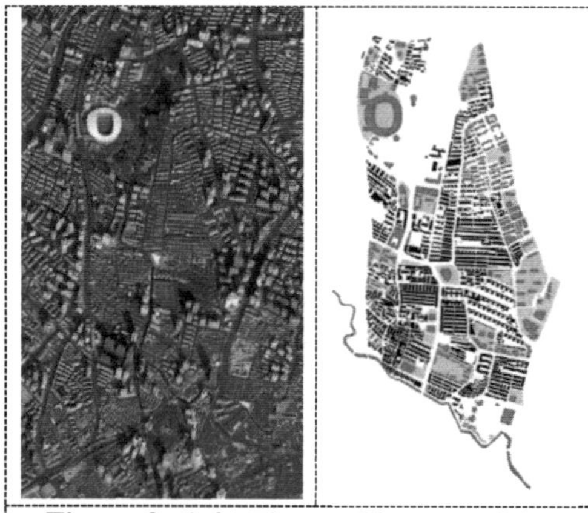

Figure 3. (left) The satellite map of the North Sichuan Road district. Source: Baidu map website; (right) The morphological analysis of the North Sichuan Road district which the red points represent the new form plots. Source: Drawing by the author.

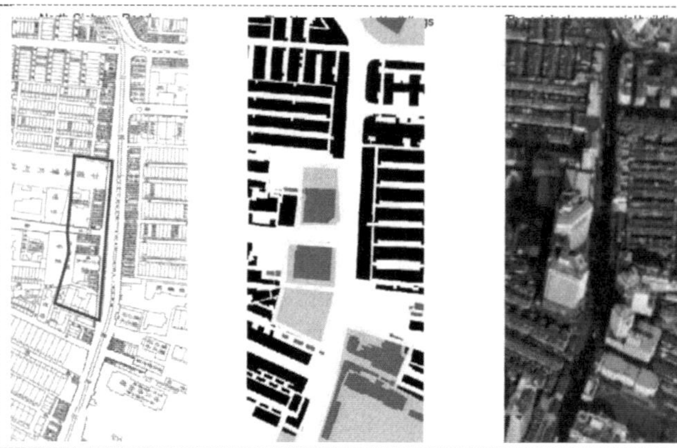

Figure 4. The insertion of the new project and the demolition of the traditional forms. Source: Drawing by the author.

3.1. Declined linkage of the public space

The traditional public structure was destroyed when the small-scale paths had been demolished (Figure 5; Figure 6). The streets and alleys, which initially linked the differential urban patterns, were cut off by the reformed squares and towers. Besides, the North Sichuan Road, which had worked as the pedestrian commercial street in this area, was transformed into main vehicle street with huge shopping malls. The current mixture use of pedestrian and vehicle had caused severe safety problem and traffic jam.

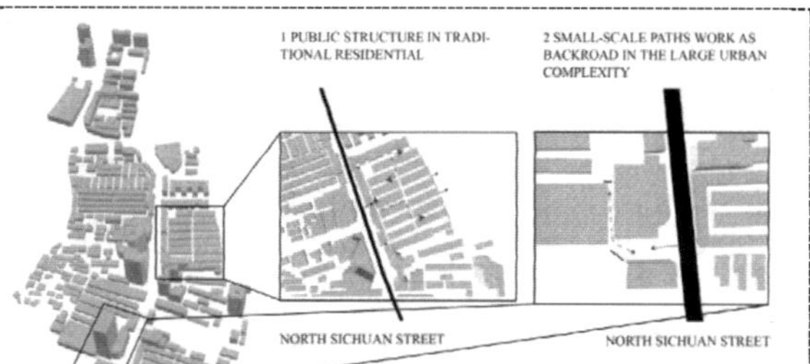

Figure 5. The differential public structure in traditional residential plots and the large urban complexity. Source: Redraw from the studio work.

Figure 6a. The traditional form. Source: Drawing by the author.

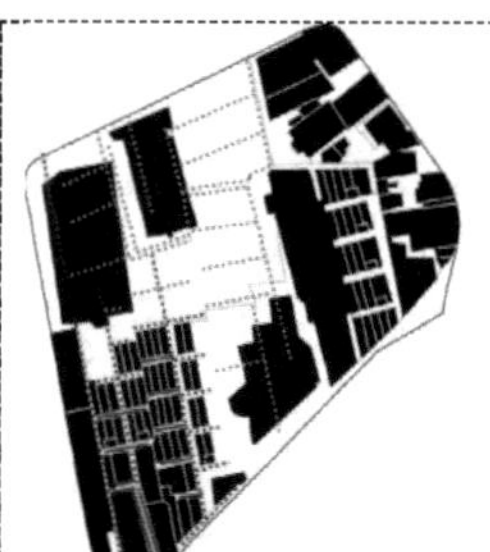

Figure 6b. The new form and the destroyed traditional public structure. Source: Drawing by the author.

3.2. Decreased interaction of the various nodes

The unrelated relationship among the present commercial function and the local community need to bring further hardship in the development of the North Sichuan Road district. It is hard for the small shops to compete with the super-malls, so they gradually closed or had been demolished. Nevertheless, these new commercial complexities were built to full-fill the consumption of the middle classes, while more than half of the residents in this district remain the low-income immigrants. Although there were still traditional residents remained, they lost most of their local commercial shops and other societal nodes such as small gardens and lanes. The various interaction nodes in this district severely decreased.

4. Rethinking the future conservation and renovation

The issue I raised here is how to rebuild the public structure in a way which can manage vehicle transportation, as well as the pedestrian communication and the historical interface. As stated by Foucault, the core mission of the conservation and regeneration of the North Sichuan Road district is to entertain the claims to the attention of local, discontinuous, disqualified, illegitimate knowledge against the so-called order (Foucault, 1980)[14]. Through the micro-renewal of local links and nodes, the connectivity of the traditional neighborhood could be repaired. (Figure 7).

4.1. Reconstruct the hierarchy of the street system

There is a hierarchy structure in current cities in which the capacity of activities improved from narrow paths to highway channels. Through the reconstruction of the order of the public space, the platform of the human behaviors in differential scales could be built. In order to increase the connectivity of each level in the North Sichuan Road district, there are three key points to improve the street capacity of the pedestrian behaviors and vehicle transportation: (1) reconstruct the hierarchy of street system; (2) increase the number of the small-scale paths; (3) improve the diversity of the sidewalk design based on specific function of the street.

4.2. Recover the typical traditional nodes

Prominent buildings and monuments provide nodes for long-lived community interactions. Thus, start with the rebuilt of these significance nodes, it is possible to trigger the interaction of societal activities and recover the episodic memory of traditional neighborhood. In the case of the North Sichuan Road district, the large squares of the new projects can be redesigned to link the remained small paths of the surrounding plots. It can be achieved through three strategies: (1) design the new projects with local morphological patterns; (2) recover the typical traditional nodes, especially the old memorial landmarks of the site;(3) Rebuilt the small-scale continuity paths to link the essential modern and traditional urban nodes.

4.3. Coordinative operation to long-term manage societal interaction

Through the coordinative operation of the large and small projects, the city area would develop new networks towards differential urbanism. The large-scale projects in the North Sichuan Road district run a single function facing similar customers which had established the intangible boundaries. However, the original local business contained functions facing differential needs throughout the whole day. Shops and offices open during the day, restaurants and opera houses open in the night, and small bars serve to late night. Thus, if these large projects can be subdivided into small functional units, and partly cooperate with the surrounding private shops to cover the various kinds of entertainments and facilities. In this case, it should be useful in attracting residents and even tourists, form new nodes and manage long-term societal interaction in the North Sichuan Road district.

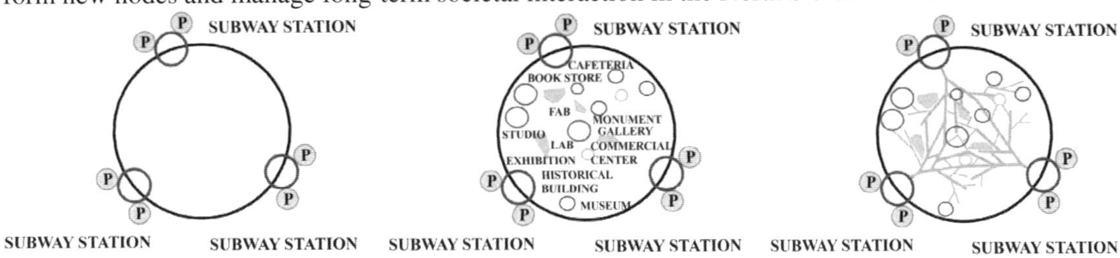

Figure 7. The conceptual approaches in reconnecting the urban network:(1) Reconstruct the hierarchy of the street system;(2) Recover the typical traditional nodes;(3) Coordinative operation to long-term manage societal interaction. Source: Redraw from the studio work. Source: TMstudio.

5. Conclusion: Rules towards a differential urbanism

To sum up, this research responds to the questions on how to revitalize the traditional district and the approaches to connect differential urban fabrics. The urban texture is a way of guiding the city to establishing order. It deals with "possible forms of the urban environment (Lynch, 1984)[15]". With the deep understanding of the existed urban fabrics, strategies on physical elements can trigger a vivid city image and social reality.

6. References

[1] Pinto J, Remesar A 2012. Public space networks as a support for urban diversity. Open House International 37(2): 15–23.
[2] Alawadi K 2017. Rethinking Dubai's urbanism: Generating sustainable form-based urban design strategies for an integrated neighborhood. Cities. 60. 353-366.
[3] Rowe C, Koetter F 1975. Collage city. Cambridge: MIT Press.
[4] Alexander C 1977. A pattern language: Towns, buildings, construction. Oxford University Press.
[5] Nikos A. S. 1998. Theory of the urban web[J]. Journal of Urban Design. 1
[6] Gosling D, Maitland R B 1984. Concepts of Urban Design. London: Academy Editions.
[7] Buchanan P 1988. What city? A plea for place in the public realm, Architectural Review, 1101. 31-41.
[8] Roberts M, Lloyd-Jones T, Nice S 1999. Place and space in the networked city: Conceptualizing the integrated metro polis. Journal of Urban Design. 4(1). 51–66.
[9] Pinto A J, Brandão A L 2015. A multi-scale approach of public space networks in the scattered city. URBAN DESIGN International, 20(3).175–194.
[10] Remali, A., Porta, S., Romice, O., Abudib, H. 2015. Street quality, life and centrality, in Vaughan, L. (ed.) Suburban urbanities: suburbs and the life of the high street. UCL Press, London.104–29.
[11] Remali, A., Porta, S. Street 2017.networks and street-blocks in the city centre of Tripoli. Urban morphology. 21 (2).161-179
[12] Carmona M, Heath T, Oc T, Tiesdell S 2003. Public Places–Urban Spaces: The Dimension of Urban Design. London: Architectural Press.
[13] Pinto, J. and Remesar, A. 2012. Public space networks as a support for urban diversity. Open House International 37(2): 15–23.
[14] Foucault 1980. Language, Counter-Memory, Practice. New York: Pantheon.
[15] Kevin L 1984. City sense and city design: writings and projects of Kevin Lynch. Cambridge, Mass: MIT Press.

IOP Conf. Series: Materials Science and Engineering **690** (2019) 012023 doi:10.1088/1757-899X/690/1/012023

Gap Acceptance for Yangon Urban and Suburban Intersections

Kyaw Nyein Nyein San[1] and Siradol Siridhara[2]

[1]Department of Civil and Environmental Engineering, Faculty of Engineering, Mahidol University, Nakhon Pathom, Thailand
[2]The Cluster of Logistics and Rail Engineering, Faculty of Engineering, Mahidol University, Nakhon Pathom, Thailand

Email: kyawnyeinsun.25@gmail.com

Abstract. Gap acceptance plays an important role in intersection design and signal warrant. This study investigates gap acceptance of drivers in Yangon, Myanmar. Three T-intersections were selected for study sites; two intersections in the urban area and the other one in Yangon suburb. As Yangon drivers drove on the right-hand side, the key movements that defined the intersection capacity would include left-turn from the major stream (LT-major), right-turn from the minor stream (RT-minor), and left-turn from the minor stream (LT-minor). Raff's method was applied to determine the gap acceptance for these three conflicting movements. The result showed that the gap acceptance for LT-minor was the longest ranging between 7 to 15 seconds. The shortest gap acceptance belonged to the RT-minor for 4 to 8 seconds. The gap acceptance at urban sites was also shorter than that of suburban in all three movements. The drivers tended to decrease their gap acceptance when the traffic volume increased. However, the relationship between the gap acceptance and traffic flow was not significant.

1. Introduction

Yangon is the largest and the most important commercial city of Myanmar. In 1941, after the British colonized time, the population of Yangon was about half a million. The number steadily increased and reached approximately 1.6 million in 1950. The population growth rate ranges from 22.2% to 22.4% annually [1]. As more rural residents continue to move to urban for various reasons, the urbanization of Yangon has been rapidly increased. Based on the current growth rates, it is expected the population would top 7 million in next 20 years. Figure 1 shows rapidly growth of Yangon population.

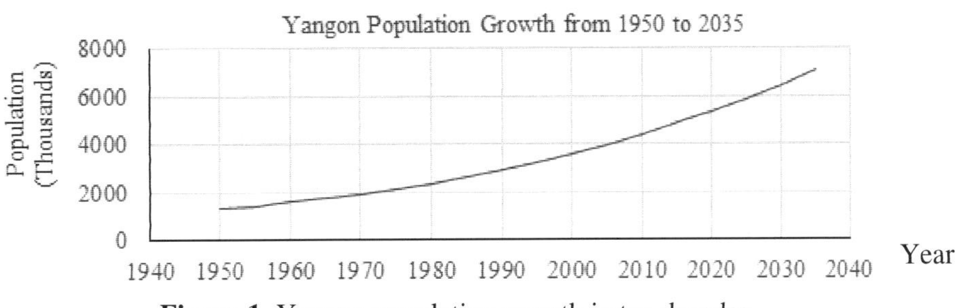

Figure 1. Yangon population growth in ten decades

Rapid urbanization and motorization have been observed along with the population growth. The city transport infrastructure has not been efficient to accommodate vehicles. There has not much improvement in city's infrastructure although the population growth increases significantly. In Yangon to date, there has not been an urban rail transit system. The railway is only the public transportation which provides high quality intercity service. Thus, the Yangon city's transportation mainly depends on the roadways. The increasing number of road vehicles induces traffic problems and increase accidents. In 2017, WHO reported that over 10,000 road accidents fatalities in Myanmar which accounts for 2.67% of total deaths by all causes [2]. This statistic took Myanmar to 80th rank in the world. The solutions to reduce traffic problems and accident risks are needed by investigating the traffic flow and intersections.

As traffic congestion grows, the drivers become more impatient and aggressive. This became the potential cause of the road accident [3]. It is widely accepted that the gap acceptance is an important characteristic of the drivers' performance to improve the road safety [4]. This paper will investigate and analyze the gap acceptance for Yangon in both urban and suburban regions. Then, it can provide a better understanding on the gap acceptance, which is based on traffic flow and drivers' behaviors. Moreover, this paper will verify whether the gap acceptance is influenced by the traffic flow or not. This finding can support the further studies for capacity analysis and intersection designs based on the traffic flow characteristics.

2. Importance of gap acceptance and its previous studies

The gap acceptance has been studied theoretically as well as empirically over the past years with various techniques. The main idea involves determination of a critical gap length. The gap acceptance is based on an assumption that drivers will use or "accept" gaps that are longer than the critical gap and "reject" those that are shorter. Determination of critical gap normally depends upon the conflicting traffic volume and the drivers' decision. A vehicle will stop and wait to enter intersections when the opportunity of large gap is present. If the driver waits too long for the opportunity and gets frustrated over heavy traffic stream, he or she might accept the short gap length to enter the intersection and risk collision. Thus, gap acceptance is an essential factor to improve road safety. Normally it is also an indicative factor for signal warrant.

It can usually be hypothesized that the gap acceptance can vary with the flow in one place to another. Different flow states may have different gap acceptance.Many techniques have been used to determine gap acceptance for various conditions; some of these are Raff's method [5], Logit and Probit models [6], Hewitt method [7], Maximum Likelihood Method [8] and Probability Equilibrium Method [9]. These critical gap estimating techniques and require the same information, including the accepted and rejected gaps. The gap is accepted if the vehicle on the minor stream can cross or/and merge into the major stream flow safely. Otherwise, the minor street drivers will reject the gap that they feel not safe enough to enter the intersection.

Many earlier researchers observed gap acceptance studies. Ashton [4] proposed the first assumption that the gap duration significantly influences traffic behavior. Normally, the accepted gaps are believed to be between 3 and 12 seconds [10,11]. Gap acceptance depends on several factors, a number of studies found that the shorter gap was accepted in the traffic with high flow [12,13]. The number of accepted gap was longer [12] and the number of shorter gap was found in front of small vehicles [14]. The drivers accepted riskier gaps around motorcycles than cars [15]. The accepted time gaps became significantly shorter in front of the bicycles [16]. The drivers accepted shorter gaps in straight movement than turning maneuver [17]. The gap acceptance for Australia rural regions in 2013 study was focused on the gap duration of drivers' turning maneuver and its influence on the gap acceptance behavior [18]. Although the gap acceptance could influence the drivers' decision and traffic safety, their findings pointed out that the drivers' types of maneuver could change the gap acceptance. This point can help the development of intersection designs. Similarly, the gap acceptance study for Yangon will support the improvement of Yangon's intersection designs.

3. Study sites and data collection

The study sites comprise three T-intersections; two of which are located in the urban areas and the other is in the suburban. Both urban intersections consist of two-lane major and two-lane minor roads. The suburban intersection includes four-lane major road and two-lane minor. These sites are in the moderately busy intersections of Yangon, and the traffic can be assumed as the intermediate flow including peak times. The city of Yangon drives on the right side. Thus, the gap acceptance for three turning movements are investigated; left-turn from major (LT-major), right-turn from minor (RT-minor) and left-turn from minor (LT-minor) maneuvers. The schematic study sites are illustrated in figure 2.

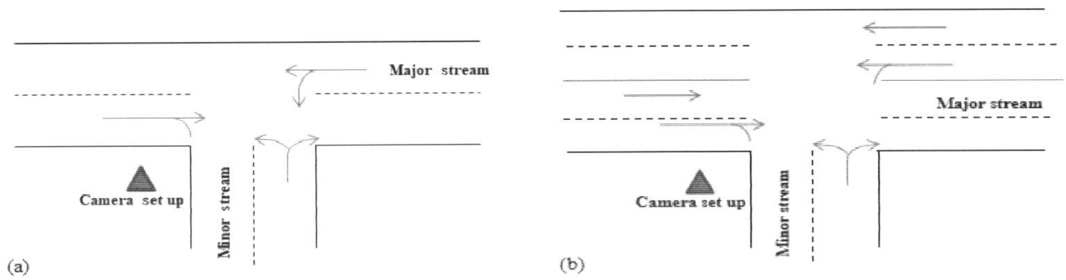

Figure 2. Plan of studied intersections (a) urban area (b) suburban area.

The camera was placed at the three study sites and the flow and gaps were retrieved from the video. The eight-hour traffic flow data were recorded in the morning, afternoon and evening peak hours for an average weekday. Traffic flows were counted in vehicles per hour from the recorded video. The arrival times of major stream vehicles at the specific point and the arrivals and departures of minor stream vehicles were measured. However, the actuate gap acceptance data could not be resulted from the site data collection of video recordings. The collected data were needed to be extracted in seconds by a vehicle-counting program to get the gaps which were required for analysis.

4. Analysis using Raff's method

The major stream traffic arrival pattern is assumed to be random, i.e., not affected by control devices. According to the results, it could be proved that the observed gaps are in general represented Pearson distribution, using a model parameter specific to the location [19]. Then, the gap acceptance is determined by Raff's method. The critical gap can be defined as the number of accepted gap less than it is equal to the number of rejected gap greater than it. In the analysis, it is noticeable that any particular driver can only accept one gap, whereas she can reject many of them at any one time especially peak hours. It can be found that there are many more rejected than accepted gaps. The cumulative distributions of accepted and rejected gaps are resulted from the observed data and used to plot three pairs of curves for three turning movements: LT-major, RT-minor and LT-minor. Then, the critical gaps can be obtained as the intersections of pairs of curves.

Table 1 shows the range of the critical gap values and it can be clearly seen the different between urban and suburban regions for each movement. In our study sites, the flow for urban area is around 300-800 veh/hr and that of suburban is between 100 and 500 veh/hr. For LT-major and RT-minor movements, the critical gap intervals are not quite different although the values are different for urban and suburban. The drivers choose to enter the intersection in shorter gaps at urban; it means the drivers may be aggressive and impatient because of traffic jam in the downtown areas. At the suburban intersection, the drivers have long acceptable gaps as the flow is not quite high, and they can choose longer gap to merge the intersection. Thus, it can be assumed that the traffic flow has the significant impact for the gap acceptance and the drivers' maneuver.

The left-turn movement from minor stream is the last priority of traffic stream and it has to be subordinate to all other streams. Thus, the drivers who want to use this maneuver need to wait longer time than others and also they need longer acceptable gaps as their maneuver cross the intersection. The acceptable gap for this movement is ranging from 7 to 11.5 sec for urban and around 9.5-15 sec for suburban. Longer gaps are accepted for both regions. It may be suggested that the drivers are patient and can wait to cross the intersection, however, those longer gap acceptance are mainly based on the traffic flow. We can prove that point by comparing the interval values of both areas. The critical gap range for urban area is found shorter than that of suburban area, as the traffic flow for urban is higher than suburban. According to the results, it can conclude that the traffic flow is inversely proportional to the minimum gap acceptance. The higher the flow in traffic stream, the lower gap values the minor street drivers will get to enter the major stream.

Table 1. Critical gap intervals for three movements in urban and suburban regions (seconds).

Movements	Urban	Suburban
LT-major	4.62 to 8.17	6.67 to 11.33
RT-minor	4.21 to 6.44	6.00 to 8.00
LT-minor	7.00 to 11.29	9.71 to 15.00

The resulting critical gap values are displaced in the scatter plots. Figure 3 shows three trend lines for three turning movements for urban area (a) and suburban area (b) respectively. In both parts (a) and (b), it is visible that the gap acceptance values decreases along with the increasing traffic flow. That shows the gap acceptance can vary with the traffic flow.

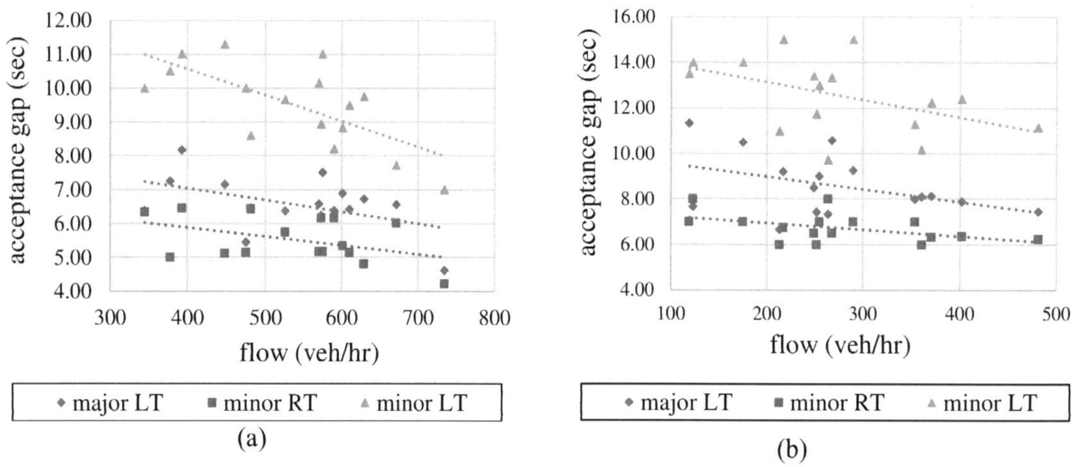

Figure 3. Critical gap interpretation for three movements in urban (a) and suburban (b) areas.

While the data show the same trends, the regression slope t-test method is used to validate this finding. Table 2 illustrates the results of linear regression analysis for both urban and suburban areas. However, all slope values are not much different from zero. This provides no evidence that the gap acceptance values vary significantly with the flow of traffic streams. The coefficient of determination; R^2 values, are around 17 to 23 % except for one urban movement. Although R^2 values are lower values, it cannot assume that the dependent variable; gap acceptance, has less actuate predictions from the traffic flow as the observation results depend on the human behavior [20]. However, the hypothesis testing with 95% level of confident gives the unexpected results. The p-values for all movements except one urban movement (LT-minor) are greater than the alpha value (0.05). P-value needs to be lower than alpha value to confirm the significant appearance [21]. This result is quite different than others. Although we can see clearly that all trend lines decrease when the traffic flow increases, it cannot be proved that the traffic flow has significant effect to the gap acceptance. It may be suggested that the relationship between the gap acceptance and traffic flow is not apparent. It cannot say, however, there is no relationship between the gap acceptance value and the traffic flow. It can be resulted that the drivers tend to decrease their gap acceptance when the traffic volume increases. Thus, urban area has lower critical gap values than suburban area.

Table 2. Regression slope test and hypothesized results.

Region	Movement	a	b	R^2	p- value	alpha
Urban	LT-major	-0.0035	8.4414	0.2259	0.0628	0.05
	RT-minor	-0.0026	6.9232	0.1822	0.0992	0.05
	LT-minor	-0.0077	13.63	0.4718	0.0033	0.05
Suburban	LT-major	-0.0055	10.083	0.1735	0.1086	0.05
	RT-minor	-0.0029	7.5339	0.2175	0.0686	0.05
	LT-minor	-0.0077	14.689	0.2278	0.0616	0.05

5. Conclusion

Critical gaps for left-turn from major stream and right-turn from minor stream (regarded as Rank 2 in traffic priority) are quite similar; 4 to 8 seconds in urban and 6 to 11 seconds in suburban areas. Left-turn movement from minor stream (Rank 4, the last priority) has the longest critical gaps. The resulted actuate gap values for urban and suburban areas of Yangon can be seen in Table 1. According to the observation results, drivers at suburban intersection have longer critical gaps than urban ones. Lighter traffic tends to result in longer gaps. The relationship between the volume and gap acceptance is not evident. This gap acceptance study has the limitation for U-turns. In some intersections with limited capacity, if U-turn is to be allowed, the gap acceptance for the U-turn maneuver must also be investigated. Care should be taken as U-turns most probably require very long gap and affect intersection capacity significantly. In the further stages, the investigation of gap acceptance behavior will be applied to signal warrant of intersection designs for Yangon. However, the U-turns condition needs to be considered in the future research for the gap acceptance of Yangon.

6. References

[1] Yangon Population 2019 (worldpopulationreview.com/world-cities/yangon-population/)
[2] World Health Rankings(https://www.worldlifeexpectancy.com/myanmar-road-traffic-accidents)
[3] Present Situation of Road Safety in Myanmar (www.unescap.org)
[4] Ashton W D 1971 Gap-acceptance problems at a traffic intersection *J.Appl. Stats.* **20** 2130-138
[5] Raff M S and Hart J W 1950 *Traffic Engineering and Control* **5** pp 255-258
[6] Kensher D A, Rose J M, Greene W H 2005 *Applied Choice Analysis: A Primer* (Cambridge)
[7] Hewitt R H 1983 *Transportation Science* **17** 187-109
[8] Troutbeck R 1992 *Research Report* (Brisbane: Queensland University of Technology) pp 92-5
[9] Wu N 2001 *Transportation Research Part B* **35** 6593-623
[10] Madanat S M, Cassidy M J and Wang M H 1994 *J. Trans. Engr.* **120** 121-26
[11] Gittis J L and Low S T 1999 *J. Trans. Engr.* **125** 3201-207
[12] Bottom C G and Ashworth R 1978 *Ergonomics* **21** 9721-734
[13] Yan X, Radwan E and Guo D 2007 *Accident Analysis and Prevention* **39** pp 843-852
[14] Mitsopoulos-Rubens E and Lenne M G 2012 *Accident Analysis and Prevention* **49** pp 86-95
[15] Chole J R, Harriet A A and Peter C 2018 *Transportation Research Part F* **58** pp 944-954
[16] Petzoldt T, Schleinitz K, Krems J F and Gehlert T 2017 *Safety Science* **92** pp 283-289
[17] Alexander L, Cheng P and et.al. 2007 *TRB 86th annual meeting* (Washington, DC)
[18] Beanland V and et.al. 2013 *Transportation Research Part F* **20** pp 80-89
[19] Nyein Nyein San K and Siridhara S 2019 Flow-headway distribution relationship: a case study of Yangon *NCCE24* (Udonthani: Thailand) **TRL050** pp 1970-1976
[20] Ortuzar J D and Willunsen L G 2011 *Modelling Transport, 4th Edition* (United Kingdom)
[21] Ross S M 2010 *Introductory Statistics, Third Edition* (University of Southern California)

IOP Conf. Series: Materials Science and Engineering **690** (2019) 012024 doi:10.1088/1757-899X/690/1/012024

Interior Living Space Preferences in the Early Housing Career of Low-Income People in DKI Jakarta, Indonesia

Antony Sihombing, Coriesta Dian Sulistiani, Rossa Turpuk Gabe, Cut Sannas Saskia and Chotib

Universitas Indonesia, Indonesia

Email: coriestadian@ui.ac.id

Abstract. DKI Jakarta has been the most prospective destination for Indonesian low-income urban migrants to find jobs. The need for affordable housing for low-income people increase along with the rise of low-income urban migrants' number. Providing houses for Indonesian low-income people has become the Indonesian government's concern as the housing providers for many years. There were unmatched between the housing government-supplied and the demands of the low-income people. Low-income people have unique preferences in choosing their residences. Moreover, the previous studies show one's preferences on housing develop over one's lifecycle, namely housing career. The preferences for living space is one of the house attribute importance that considered in choosing a residence. This paper will discuss the low-income people preferences on the living spaces and the changes of these preferences over their lifecycle, especially in their early housing career. The research method used distribution of living space preference changes by selected independent variables and bivariate statistical analysis with Pearson Contingency Coefficient. The study took five of the most populated districts in DKI Jakarta and used snowball sampling on 420 respondents. The paper discussion only focuses on 369 of 420 respondents who have at least 2-residence-career, while the other respondents have never moved out from their first residence. This study found that the majority of respondents consider living space when choosing their residence, and almost half of the preferences change along with the changes in the lifecycle. The results of this study are expected to be a consideration for the Indonesian government in providing affordable housing for Indonesian low-income people.

1. Introduction

Indonesia government ensure its citizens for affordable housing. This is as regulated as on the Law of the Republic of Indonesia Number 1 the year 2011 concerning Housintableg and Settlements. Based on the law, providing adequate and affordable housing for the low-income people and understanding the consideration low-income people made in choosing a residence become essential concerns for Indonesian government as achieving the aims of Indonesian Government National Program, namely One-million House. Sylvia et al. [1] differentiate between the preferences and choices. Preference is a subjective desire that comes from the internal one's self, while choice is the meeting point between the individual preferences and external-self conditions, such as the housing availability, housing prices, and housing regulations. So, one's choice of a residence influenced by one's preferences. Opoku and Abdul Muhmin [2] stated that choosing of a residence is not only influenced by one's preferences but also the house attribute importance. The house attribute importance influences the housing choice from intrinsic housing attributes such as cost, size, and interior living space through extrinsic attributes such as exterior design and exterior space to neighborhood and other locational factors.

Jim and Chen conducted a study that shows South China residents rate internal environment elements of the house as more important than external [3]. The residents consider an internal environment as a place where the residents actually live. From the Opoku and Abdul Muhmin' result study in Saudi Arabia, the residents also consider the intrinsic attributes have significant meaning [2]. In everyday life, internal living space of a house can serve a symbolic function, providing clues for others about the identity, social class, aesthetic preferences, personality traits, and personal histories of the residents [4]. It may show the social demographic conditions of the owner. However, people grow, change, and evolve. One's lifecycle changing can affect one's preference for the intrinsic attributes in housing choice. It is as stated as Mulder & Hooimeijer' argument that housing preference of an individual or family changes as the life-changing [5]. The changes can also be influenced by events or life plans such as marriage plan, career plan, work location, and other else [6]. Kok explained the individual or family's life course is dynamic and complex; therefore, housing preference will be dynamic and unpredictable [7]. Hence, individual or family preference is divided into durations that caused by events in every life stages of households [8].

Published academic studies of consumers' housing preferences and related issues have been conducted in marketing perspectives with medium-high income people as its respondents. The focus of this paper is on the low-income segment in their early housing career. Living space availability that will support individual life needs can be one of their consideration in choosing a residence, and the preferences can differ from one to another stage of their life. Understanding the low-income people demand on living space of a house can be a consideration for the government to make an appropriate housing design so that it can maximize its occupancy.

2. Aims of the study

Various studies prove that the mismatch of housing design with social, economic, and cultural conditions are the leading causes of the government' housing programs for the Indonesian low-income people failure [9]. Indonesian government believes that affordability of rental or purchase prices is the only factor that determines the preference of low-income people in choosing a residence. Whereas in choosing a residence, low-income people have unique preferences and have a housing career that related to their needs and social, economic, and cultural conditions. Every residence that has been chosen has different backgrounds of preference that related to life needs and social, economic, and cultural conditions at that time. Therefore, these preferences continue to change as the low-income people life cycle changes. Unfortunately, the lack of understanding of the government regarding the changing factors that influence the preference of the low-income people has caused various housing programs has not achieved the expected results. This study aims to identify and understand changing factors that focus on living space that influence the selection of residence in the early housing career of low-income people in Jakarta. The results of this study are expected to be a consideration for low income-income people housing providers in designing adequate living space for affordable housing.

3. Methodology

This research is located in the Indonesian Capital City, DKI Jakarta Province. As the capital city, Jakarta becomes the prospective destination for Indonesian urban migrants to find jobs. This condition makes Jakarta has the highest migration rates among other cities in Indonesia. There are 5 selected most populated villages namely Kampong Kalibaru (Cilincing Subdistrict, North Jakarta); Kampong Menteng (Tebet Subdistrict, South Jakarta); Kampong Melayu (Jatinegara Subdistrict, East Jakarta); Kampong Tanah Tinggi (Johar Baru Subdistrict, Central Jakarta); and Kampong Kapok (Cengkareng Subdistrict, West Jakarta). Data collection uses a structured questionnaire at 50 households with a category of age over 15 years. Distribution of questionnaires in each village was carried out by snowball sampling technique so that 420 respondents were collected. Each respondent has a different living career with diverse reasons for housing choice. About 234 respondents had a one-time residential career, 111 respondents who had a two-times residential career, 23 respondents who had a three-times residential career, one respondent who had a four-times residential career, and 51 respondents who never experienced moving out from their first house. This study focuses on 369 of 420 respondents who have at least a 2-residence-career. This study aims to determine the relationship between living space preferences and the life-changing in the early housing career of low-income people. The analysis was carried out with a quantitative approach by descriptive and inferential analysis. The descriptive analysis used cross-tabulation between the dependent variable and selected independent variables. The inferential analysis used chi-square method to see the magnitude association between dependent and selected independent variables. Calculation of the correlation magnitude between selected independent variables and dependent variables shows the low-income people general information about their living space preferences in Jakarta.

4. Results and discussions

4.1. Data Collection

Data was collected from 369 respondents spread in 5 of the most populated villages in DKI Jakarta. The distribution of respondents in these five villages can be seen in table 1. Most respondents were in Kampung Melayu with 82 respondents and the fewest in Tanah Tinggi as many as 63 respondents.

Table 1. Distribution of Respondents of 5 locations in 5 Districts of DKI Jakarta Province

District	Total	Percentage
Tanah Tinggi	63	17.07%
Kapuk	69	18.70%
Kalibaru	76	20.60%
Menteng Dalam	79	21.41%
Kampung Melayu	82	22.22%
TOTAL	**369**	**100.00%**

The chosen living space preference of the respondents are a living room, bathroom, bedrooms, additional bedrooms, kitchen, laundry room, storage room, garage, terrace, other, and some of the respondents did not choose any rooms as their interior living space preference. Based on data collected, almost one-third of the respondent has no preference for interior living space. To simplified the collected responds of the questionnaire, the eleven preferences options are simplified into four categories, namely:

- Primary rooms (bedrooms, kitchen, bathrooms)
- Secondary rooms (living room, family room)
- Tertiary rooms (garage, storage room, terrace, laundry room, additional bedrooms, other)
- No preference

Based on data collected, almost one-third of the respondent has no preference for interior living space. However, these interior living space preferences changed as the respondent moved to their second residence, as shown in Table 2.

Table 2. Living Space Preference in the Early Housing Career Low-Income People in DKI Jakarta (re-coded version)

Preference for living space element	first house		second house	
	n	%	n	%
Primary rooms	88	23.85%	103	27.91%
Secondary rooms	91	24.66%	99	26.83%
Tertiary rooms	44	11.92%	59	15.99%
No preference	146	39.57%	108	29.27%
TOTAL	**369**	**100.00%**	**369**	**100.00%**

Based on the data analysis, the changes in the preferences and the socio-demographic condition of the respondents from the 1st residence to 2nd residence shown in table 2. Based on this result, an in-depth analysis was conducted to see what changes that have the most significant association to the interior living space preference changes.

4.2. Analyses and Result

Facilitating multivariate statistical analysis with Pearson Contingency Coefficient, the changes in interior living preferences are simplified into four categories, namely:

- No preference: both on the 1st residence and 2nd residence, the respondent has no preference
- Downgrade: on the 1st residence, the respondent has upper-level rooms preference and change it into lower level rooms on the 2nd residence (e.g., tertiary rooms to secondary rooms, secondary rooms to primary rooms)
- Steady: both on the 1st residence and 2nd residence, the respondent has the same level rooms preference (e.g., tertiary rooms to tertiary rooms, secondary rooms to secondary rooms, primary rooms to primary rooms)
- Upgrade: on the 1st residence, the respondent has lower level rooms preference and change it into upper-level rooms on the 2nd residence (e.g., primary rooms to secondary rooms, secondary rooms to tertiary rooms)

Based on Table 3, the respondents who have no preference in the 1st residence and 2nd residence mostly are the respondent who stayed out of Jakarta, lived less than 20 years on the 1st residence and more than 20 years on the 2nd residence, has steady occupation status, turned to be not married on the 2nd residence, were the head of the family on the 1st residence but turned to be not the head of the family on the 2nd residence, and upgraded the ownership status of the residence. The respondents who downgraded their preference mostly are the respondent who moved out of Jakarta, lived more than 20 years on the 1st residence and less than 20 years on the 2nd residence, has steady occupation status, were not married on the 1st residence, were the head of the family on the 1st residence but turned to be not the head of the family on the 2nd residence, and downgraded the ownership status of the residence. The respondents who have steady preference are mostly the respondent who moved into Jakarta, lived more than 20 years on the 1st residence and less than 20 years on the 2nd residence, has upgraded occupation status, stayed not married on the 1st residence and the 2nd residence, were the head of the family both on the 1st residence and 2nd residence, and upgraded the ownership status of the residence. Meanwhile, the respondents who have upgraded preference mostly are the respondent who stayed in Jakarta, lived more than 20 years both on the 1st residence and the 2nd residence, has downgraded occupation status, turned to be married on the 2nd residence, were not the head of the family on the 1st residence but turned to be the head of the family on the 2nd residence, but downgraded the ownership status of the residence.

Table 3. Distribution of Living Space Preference Changes by Selected Independent Variables

Independent Variables	**Variable Category**	**Living Space Preference Changes**				**Total**	
		No preference	**Downgrade**	**Steady**	**Upgrade**	**Count**	**Row %**
		Row %	**Row %**	**Row %**	**Row %**		
Location towards DKI Jakarta changes	Moved into DKI Jakarta	30.2 %	16.9 %	**29.2 %**	23.4 %	154	100%
	Moved out of DKI Jakarta	0 %	**25 %**	25 %	50 %	8	100%
	Stayed in DKI Jakarta	25.2 %	18 %	28.6 %	**28.2 %**	206	100%
	Stayed out of DKI Jakarta	**100 %**	0 %	0 %	0 %	1	100%

Period of the house occupation changes	< 20 years to > 20 years	**58.3 %**	0 %	18.8 %	22.9 %	48	100%
	> 20 years to < 20 years	2 %	**37.2 %**	**37.9 %**	22.8 %	145	100%
	< 20 years to < 20 years	46.3 %	1.3 %	24.2 %	28.2 %	149	100%
	> 20 years to > 20 years	0 %	33.3 %	22.2 %	**44.4 %**	27	100%
Occupation status changes	Downgrade	13.3 %	13.3 %	26.7 %	**46.7 %**	30	100%
	Steady	**30.8 %**	**20.2 %**	26.3 %	22.7 %	198	100%
	Upgrade	24.8 %	14.9 %	**32.6 %**	27.7 %	141	100%
Marital status changes	Married-married	27.9 %	10.8 %	36.9 %	24.3 %	111	100%
	Married-not married	**77.8 %**	0 %	0 %	22.2 %	9	100%
	Not married-married	24.3 %	**21.3 %**	23.3 %	**31.2 %**	202	100%
	Not married-not married	27.7 %	**21.3 %**	**38.2 %**	12.8 %	47	100%
Relation with the head of the family	Head of the family-head of the family	31 %	12 %	**34.5 %**	22.4 %	58	100%
	Head of the family-non-head of the family	**75 %**	**25 %**	0 %	0 %	4	100%
	Non-head of the family-head of the family	26.9 %	17.6 %	23.5 %	**31.9 %**	119	100%
	Non-head of the family-non-head of the family	25 %	19.1 %	30.9 %	25 %	188	100%
Ownership status changes	Downgrade	23.8 %	**23.8 %**	19 %	**33.3 %**	21	100%
	Steady	**36.2 %**	20.8 %	26.9 %	16.2 %	130	100%
	Upgrade	22 %	15.1 %	**30.7 %**	32.1 %	218	100%

Based on the bivariate statistical analysis with Pearson Contingency Coefficient on Table 4 and hypothesis tested by Chi-square method, the association between the independent variables and the interior living space preference that has the most significant value is the period of the house occupation changes as much as 0.546.

Table 4. Pearson Contingency Coefficient by Selected Independent Variables

Independent variables	**Contingency Coefficient**	
	Value	Approx Sig*
Location towards DKI Jakarta changes	.151	.476
Period of the house occupation changes	.546	.000
Occupation status changes	.177	.063
Marital status changes	.273	.001
Relation with the head of the family	.174	.240
Ownership status changes	.212	.008

*Notes: significant at α=10%
Source: Data analyzed by authors

The bivariate statistical analysis shows a more accurate result of the association between the independent variables and the interior living space preference that has the most significant value is the period of the house occupation changes than the analysis with distribution of living space preference changes by selected independent variables. Based on the bivariate statistical analysis, the result shows that living space preference changes have association significantly with the period of the house occupation changes, occupation status changes, marital status changes, and ownership status changes.

5. Conclusion

The interior living space preference has a significant association with the period of the house occupation changes. This paper focuses on the early housing career of low-income people in five kampongs in Jakarta, Indonesia. A further examination needs to be followed to have a deeper understanding of the low-income people's preference in choosing the interior living space of their residence. The result of this study is different from several previous studies that have conducted before. In past studies, low-income people's preferences of interior living space influenced by a socio-demographic aspect such as income, education, and marital status. However, in this study, it was found that the period of the house occupation had the most influence on low-income people on interior living space.

6. References

[1] Sylvia, J., Jansen, and Coolen, R. W. (2011). The Measurement and Analysis of Housing Preference and Choice. Heidelberg: Springer.
[2] Opoku, R. A., and Abdul-Muhmin, A. G. (2010). Housing preferences and attribute importance among low-income consumers in Saudi Arabia. Habitat International, volume 34, issue 2 (April 2010), 219-227. DOI= https://doi.org/10.1016/j.habitatint.2009.09.006
[3] Jim, C. Y. and Chen, Y. W. 2006. Consumption preferences and environmental externalities: A hedonic analysis of the housing market in Guangzhou. Geoforum, volume 38, issue 2 (March 2007), 414-431. DOI= https://doi.org/10.1016/j.geoforum.2006.10.002.
[4] Cupchik, G.C., Ritterfield, U., and Levin, J. 2003. Incidental Learning of Features from Interior Living Spaces. Journal of Environmental Psychology, volume 23, issue 2 (June 2003), 189-197. DOI= https://doi.org/10.1016/S0272-4944(02)00103-2.
[5] Mulder, C. H., Hooimeijer, P. 1995. Moving into owner-occupation. Journal of Housing and the Built Environment 10: 5–25.
[6] Clark, W. A. V., Dieleman, F. M. 1996. Households and housing: choice and outcomes in the housing market. New Brunswick, NJ: CUPR Press.
[7] Kok, J. 2007. Principles and prospects of the life course paradigm. Annales de Démographie Historique 1: 203–230.
[8] Clark, W. A. V., Deurloo, M. C., Dieleman, F. M. 2003. Housing careers in the United States, 1968–93: Modelling the sequencing of housing states. Urban Studies 40: 143–160.
[9] Sihombing, A., Adianto, J., Hasan, C., Gabe, R. T., Sulistiani, C. D., & Saskia, C. S. (2018). Preference of Location Selection at the First Residence: Study in Several Kampong in DKI Jakarta, Indonesia. Presented in ICETsAS. Lampung (unpublished seminar proceedings).

Acknowledgment

Financial support from Ministry of Research, Technology and Higher Education (Kementerian Riset, Teknologi dan Pendidikan Tinggi) of the Republic of Indonesia in Penelitian Dasar Unggulan Perguruan Tinggi grants year of 2018 (Grant No. 396/UN2.R3.1/HKP05.00/2018) is gratefully acknowledged.

AUTHOR INDEX

Institute of Physics
Dirac House, Temple Back
Bristol BS1 6BE UK

ISSN: 1757-8981
ISBN 978-1-7138-0982-1

CURRAN ASSOCIATES INC.
proceedings.com